Biosystems & Biorobotics

Volume 3

Series Editor

E. Guglielmelli, Roma, Italy

For further volumes:
http://www.springer.com/series/10421

Giovanni Vecchiato · Patrizia Cherubino
Arianna Trettel · Fabio Babiloni

Neuroelectrical Brain Imaging Tools for the Study of the Efficacy of TV Advertising Stimuli and Their Application to Neuromarketing

Springer

Giovanni Vecchiato
Dept. of Physiology and Pharmacology
University of Rome Sapienza
Rome
Italy

Patrizia Cherubino
BrainSigns srl.
Rome
Italy

Arianna Trettel
BrainSigns srl.
Rome
Italy

Fabio Babiloni
Dept. of Physiology and Pharmacology
University of Rome Sapienza
Rome
Italy

Additional material to this book can be downloaded from http://extras.springer.com

ISSN 2195-3562 ISSN 2195-3570 (electronic)
ISBN 978-3-642-38063-1 ISBN 978-3-642-38064-8 (eBook)
DOI 10.1007/978-3-642-38064-8
Springer Heidelberg New York Dordrecht London

Library of Congress Control Number: 2013936729

Printed on acid-free paper

Springer is part of Springer Science+Business Media (www.springer.com)

He who loves practice without theory is like the sailor who boards ship without a rudder and compass and never knows where he may cast

Leonardo Da Vinci (1452-1519)

Foreword

Neuromarketing: A Gym for Intellectual Humility

The job of a researcher in the social sciences, which economy, in all its declination, including marketing, belongs to, is both simple and difficult

The simplicity lies in the methods. Normally, the methods and techniques that social researchers have to manage to find the answers they are searching for are relatively easy to learn and apply. Someone may object at this point and say that this is not always true, for example quantitative modeling is not easy at all. We could agree on such objection.

However, even in those cases, unless the aim of the modeling strategy is to refuse the reality, which is observed to behave according to the working hypotheses, sooner or later, somehow, an understandable results may appear. You could spin the data, change the axes of the variables, look the data from different perspectives, even torture the data and at the end a result come out.

Now we come to the difficult part of the story. That is, what are the questions to be asked in research?

This is the main challenge of our job as scholars of social issues, as well as its specific cultural depth. The novelist Milan Kundera opened up his own novel with a famous quotation from the German philosopher Heidegger: "The essence of man is shaped like a question." This is a perfect snapshot of humans, the only animal capable of thinking (at least to our knowledge, who knows indeed with absolute certainty if other primates are able to do that?).

Find a reasonable, relevant and significant question is the essence of science and our existence as scholars depends on this.

Find your question. That is the hard part

The trap of "waste of intelligence" in economic research is well-oiled, hidden and always present along our way of scholars. It is a trap that lurks in a dark cavern where methods and research questions are intertwined. When we browse the scientific journals we often perceive one or more of the following feelings:

- The title is too difficult to understand;
- Some scientific papers seems to be nothing more than a statistical exercises. The market is used only as an excuse, because it offers a dataset and a framework of assumptions on which statistical methods can be applied at the

purpose of generating a study and answer questions that appears to be relevant and meaningful;
- The conclusions of the work says nothing more than the following "when it rains, we get wet". You could do that without all that work.

Finding the "right" research questions is definitely a big problem, and not only in marketing studies. In this respect you may have already heard about the organization *Improbable Research* http://www.improbable.com/ and the IgNobel Prize, which is awarded every year to scientific research conducted seriously and adequately disclosed, although often irrelevant for the choice of the research questions. It should be noted that the initiative is actually very serious: the ceremony takes place at Harvard and the winners are then invited to give a lecture at MIT!

Finding the right question: this is the core of our job as researchers

In marketing the key issue is represented by the word “relevance”, which involves a mix consisting of:

1. Culture
2. Common sense
3. A psychological problem that grips the community of the economists

Let's see one by one these three aspects, for which we will propose considerations on neuromarketing arising from the pages of this good book.

1 Culture

The culture to which we refer here is not the "economic" one, but that knowledge, large and branched, of human affairs, which alone allows us to be full owners of our thinking. Culture is not a thing but a way of seeing things.

In our profession the generation of a personal culture is the goal to which we all must address each of our daily effort. It is a job without end. Nor is it assured, on the contrary, it is difficult. "The roots of education are bitter, but its fruit is sweet," Aristotle used to say. Each step forward increases our insecurity since it makes us even more aware of our gaps of knowledge. Already 2500 years ago Socrate was able to describe such uncomfortable feeling with his famous words: "on my own, or Athenians, the more I know the more I realize do not know."

In short, here we face doubts and problems always experienced by humans.

In those respects, the essential requirement, essential because it produces good research, is therefore, first and foremost, a personal culture: a culture that is necessarily divided into several fields (do we mean multidisciplinary?), aware of its "natural" inadequacy and therefore always cautious in his exegesis, and humble.

We have just been faced with the first relevant aspect of neuroeconomic and neuromarketing studies.

Cultural insecurity is probably the prevailing feeling of an economic researcher when he ventures into the territories of the Mind and Brain. Such researcher is constantly faced with the shallowness of his knowledge on the matter and is afraid of saying embarrassing banalities and / or inaccuracies. It is true that studies of consumer behavior have always stimulated marketing scholars to enter into new cultural and scientific areas.

Thanks to this is a fact ascertained and taken over by marketing studies, now that "economic behavior is determined not only by the cultural and social context in which the individual developed, but also by the genetic component due to the evolution of the species (Lugli, 2012). It is also true that the awareness of the fluid nature of the decision-making process of the individual has been abiding city in the patterns of economic theory, as demonstrated by the recognition of the Nobel Prize to Daniel Kahneman and Amos Tversky in 2002. Actually economic researchers have been working for years to solve research questions around the themes of mind and brain, but their knowledge about these themes appeared weak being always marginal to the focus of their epistemological discipline.

On the other hand, researchers that are able to hand the technical tools of neuroscience, which read for example biometric measurements, and that are familiar with spatial and functional brain mapping suffer the same "weakness" in the front of "pure" marketing research questions.

Neuromarketing becomes therefore an useful cultural gym where the scientific skills of both economic researchers and neuroscientists can be trained, thus developing the knowledge of all. We do not know yet whether Neuromarketing will be able to generate lasting fruits. Surely it is and will be an interesting field of knowledge for any open-minded and curious person, and not just for "marketers excited about neuroelectrical brain imaging", as the authors nicely explained in this book (page 7).

2 Common Sense

The vertical knowledge of a topic, no matter how deep this may be, is not culture, but simply specialist learning: you just need a little to find meaningful and relevant research questions, if these are not against the common sense. According to a famous Italian encyclopedia (Enciclopedia Treccani, 2011), common sense regards the "natural ability, instinctively, to judge righteously, especially in view of the practical needs" . The dilemma becomes then clear:

The identification of the research question has to deal (also) with its practical application or research has to be "pure", enjoying the freedom to break free from the shackles of the practice, so that they can dissolve in the prairies of free thought? (Kildulf et al, 2011):

To respond to this question, we consider that in management (and in marketing as well) the practical or operational character of the research can obviously vary, and the relevance of a search has to be read in conjunction with the ability of the proposed research to renew the thought patterns of the discipline (Figure 1). We can then turn to the logic array represented in Figure 1 and identify in it four types of studies in search-economic management. The vertical axis concerns the epistemology of the research, that is, its ability to approach the truth of things, the horizontal axes relates instead to the ontology, i.e. the ability of a study to represent reality and to give us tools to modify it.

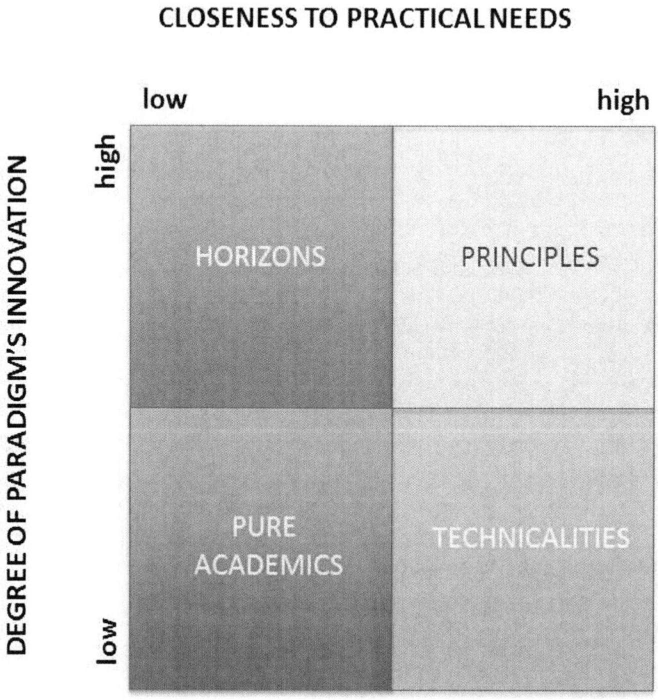

Fig. 1 An array of economic and management studies.

We will see where and how neuromarketing, which has now fully entered the scientific heritage of the discipline (Eser et al., 2011), will be linked with the conditions that arise within the matrix:

- Technicalities: a research that focuses on the practical and concrete significance of its results and their direct applicability, *mutatis mutandis*, to the most different operating environments (markets, firms, sectors). The fundamental paradigms of matter are not altered by this research, but rather up-born for discrete and incremental adjustments. Neuromarketing belongs today to the first of this important line of research. The authors of this book clearly identified the interest of marketing managers in the use of neuroscience to optimize classical instruments, such as advertising, to accelerate the return on investment, and to increase the awareness of consumers towards both brands and products, which is the ultimate interest of researchers in this field (pp. 7 et seq.) The interest of marketers for neuromarketing is vibrant and feeds also that certain "technological enthusiasm" that characterizes this first part of the twenty-first century. However, this ardor must be accompanied by concern and caution and the following points should be kept in mind: (i) how far it is permissible to push the research on the brain and is there a boundary beyond which the scientific curiosity becomes dangerous for the individual? (ii) what is the limit of heuristic experiments conducted in the laboratory by surrounding the subject with measuring instruments? Such a situation is surely not neutral with respect to the emotional state of the person in the experiment. The promises of neuromarketing are certainly fascinating, however a superior attention is needed in performing those studies, as the authors demonstrate, through all the pages of this book.

- Principles: a research that explores and measures the ravines of new areas of observation, that aims to build an architecture capable of representing knowledge, and of explaining and governing a certain portion of reality. Neuromarketing promises to make a significant advance in knowledge related to consumer behavior. At this point it seems that its contribution lies in the scientific demonstration, through precise biometric measurements, of the nature of awareness (cognitive mind) and unconscious (emotional mind) of the shares of consumers. By using the force of numbers, neuromarketing has permanently dismantled an idea that marked and accompanied economists for decades, that is the idea of a rational and linear *homo economicus*. Future neuromarketing research will probably continue to profoundly change the way in which we view and consider the complex relationships between suppliers and targets, thus modifying even more the theoretical paradigms of the discipline.
- Visions (horizons): a research produces a new thought by staring at different modes of reasoning in relation to disciplinary practices. It is a quest that is launched into an unexplored space in search of new interpretations, following the inadequacy of the old paradigms to interpret the new facts of economic reality (Kuhn et al., 2010).

The secret hope of all scholars in the economy field is that neuroscience may open new horizons in the economic thought and practice, which all fields of human endeavor may benefit from. Scientific fields related to marketing surely shares the same expectation, still the more practical nature of marketing makes marketing managers eager to obtain quick and tangible outcomes from neuroscience.

- Ruminations (Pure academics): a research that is completely self-referential in the assumptions, methods and results, as well as in communication circuits, that has neither a present nor a prospective interest in the practical relevance of its own results, and that has a limited ability of innovation.

To conclude, common sense is an important attitude of mind, which does not discriminate between good and bad research, or between theory and practice. To use a metaphor from seamanship, common sense is like a compass: alone it is not sufficient to trace the route, if not accompanied by nautical charts. In our case the nautical charts are represented by the desire to renew self- and other people´s knowledge, a desire that must always accompany the researcher.

3 The Psychological Problem

The issue of research questions is enriched by another aspect, which is less superficial but nonetheless substantial and that relates to a sense of inferiority of "applied" economics to the most scientifically accredited economy (Baccarani and Golinelli, 2011), against which some management theorists (and, among them, those involved in the marketing science) warn.

In fact, a sense of inferiority is felt by all economic sciences with respect to so called scientific or "exact" disciplines (Figure 2). The level of scientific accreditation and consequently of social reputation of the latter is indeed very high. The obscure nature of their languages, which becomes evident in sentences like "bacteria do not possess the enzymes of eukaryotic transcription", of their methods (what is a particle accelerator?) and definitions ("What is a nucleotide?") put the corresponding scientific community on a sort of superior position that remains, however, inaccessible to most people.

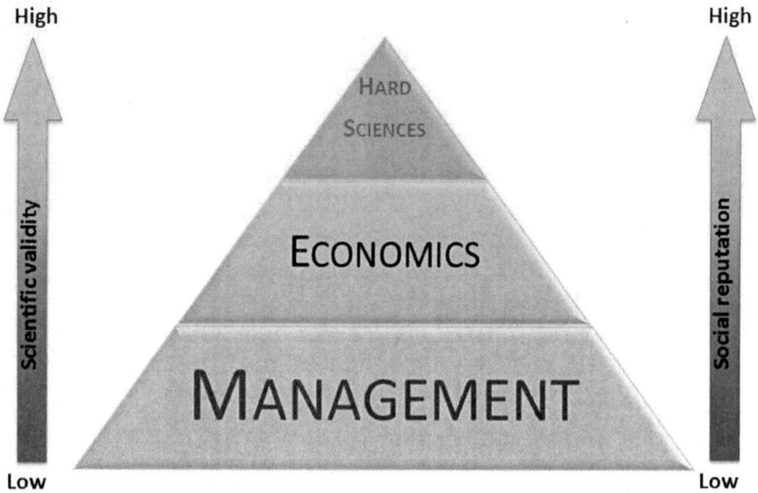

Fig. 2 The Pyramid of the sense of inferiority.

Economists suffer of the lower ranking of their discipline and feel inferior to exact scientists. They feel, so to say, as strangers that are invited to an exclusive party of English noblemen in Oxfordshire and have neither familiarity with the label of the situation nor a fluent English. This may explain why those social scientists started to make their job less and less accessible, more and more scientific, for example by introducing sophisticated quantitative methods and by using a more complicate language. The interest in neurosciences of both "pure" and "applied" social scientists may be seen as part of a process of giving scientific dignity to their discipline through the use of exact methods, languages and definitions. It is still unclear whether those efforts will pay off.

4 A Conclusion

In only one hundred years of history marketing has been able to accomplish extraordinary achievements resulting in cultural, intellectual and technical cooperation. It is now evolving from his primordial nature of "technique" in a "system of thought and action". Originally intended as "technical and sales policies" when it was born, marketing encountered during its existence many different fields and made of each one of them an original interpretation: firstly it met management, secondly it met strategy and start putting the customer on a central position, then, in recent times, it met digital science giving rise to digital marketing. The centrality of the customer and of the individual have always been a constant element in epistemological and operational disciplines. As a consequence, scholars and practitioner in marketing have come in contact with many non-economic constructs, like attitudes and personalities, lifestyles and social groups of reference.

This book is surely not the first on its subject, but it is certainly one of the most significant contribution to the field so far since it filled in an important piece in the puzzle of neuromarketing knowledge. Vecchiato et al. chose to answer the fundamental question: *how could marketing science and investment choices benefit from the analysis of brain activity in response to advertising stimuli?*

The authors of the volume are serious scholars as it could be understood from the first pages of the book, where they call for caution in the use and interpretation of those techniques and say "brain imaging techniques, applied to human decision-making mechanism, could be adopted to corroborate the results obtained by traditional techniques "(page 1) and "the use of brain imaging can, in a near future, be placed side-by-side to classical tests today largely used in the marketing sciences"(page 3). Moreover, since they properly recognize that the artificial settings of a laboratory may have an influence on the results, they aimed "to recreate, as much as possible, a naturalistic approach to the task in which the observer is viewing the TV screen without any particular goal in mind "(page 4).

It is difficult to convey the answer to the important research questions that the authors have described along the book in just a few words. Nor would it be elegant and gentle to simplify too much the complexity of this work, which involves reflection, technology and culture. With this book, Vecchiato et al. seem, however, to tell consumers the following "Do not worry, guys. Those techniques are still by no means a kind of weapons that industries may use to induce you to buy their products. " On the other side, they seem to say to managers "We are on the right track. We begin to know enough to help you optimize the architecture of advertisements, meant that our work will not help you to sell refrigerators to Eskimos, but rather to make TV advertising shorter of the old thirty seconds, and therefore cheaper."

It is definitely worth reading this book, dealing with a subject that is both fascinating and timely. This work is indeed an excellent and solid contribution from serious researchers, whose scientific commitment and intellectual humility make them great scholars.

Rome, 1.03.2013

Alberto Mattiacci
Full professor of Economy and Management
Sapienza University of Rome

References

Baccarani, C., Golinelli, G.M.: From the Editors: The Wind of Change. Sinergie 86, 9 (2011)

Bauman, Z.: Living on Borrowed Time: Conversations with Citlali Rovi-rosa-Madrazo, Cambridge (2009)

Eser, Z., Isin, F.B., Tolon, M.: Perceptions of marketing academics, neurologists and marketing professionals about neuromarketing. Journal of Marketing Management 27(7-8), 854–868

Kilduf, M., Mehra, A., Dunn, M.B.: From Blue Sky Research to Problem Solving: A Philosophy of Science Theory of New Knowledge Production. Academy of Management Review 36(2), 297–317 (2011)

Kundera, M.: Life is elsewhere. Harper Perennial (2000)

Lehmann, D., McAlister, L., Staelin, R.: Sophistication of research in marketing. Journal of Marketing 75, 155–165 (2011)

Lugli, G.: Neuroshopping. Come e perchè acquistiamo, Apogeo, Milano (2012)

Plato: Apologia Di Socrate, Laterza, Bari (2000)

Russo, L.: Segmenti e bastoncini. Universale Economica Feltrinelli, Milano (2000)

Treccani: Enciclopedia Italiana di Scienze, Lettere ed Arti, Istituto della Enciclopedia Italiana, Roma, Italia (2011)

Contents

Chapter 1
How Marketing Meets Neuroscience

The issue of how we make, and should make, decisions and judgments made think philosophers for hundreds of years and kept alive some disciplines such as philosophy and some branches of psychology. A recent approach, known as neuroeconomy, suggested to merge some ideas and scientific discoveries coming from psychology, neuroscience and economy fields as an attempt to accurately specify the models of decision-making of human mind. The rationale to integrate such as different disciplines comes from the thought that human behavior in financial field appears often irrational, as human choices at the casinos green tables, at betting offices, or also at Stock Exchanges, testify. The discipline of neuroeconomy is defined as "the application of neuroscientific methodologies for the analysis and the knowledge of interesting human behaviors in the economy field". The reshaping of the economy field is not a new event. In fact, the boundaries of the economic studies are continuously remodeled by the improvements that mathematics or simulation sciences constantly hold. In such a case, instead, the development in the understanding of human behavior rose new issues, by means of the methodologies of analysis that the cerebral activity provides, and create the convergence of different disciplines in a new area of the scientific research, such as the neuroeconomy.

"I know that half of my money spent in advertising is wasted, but I don't know which is such half", are the words that John Wanamaker, who built the first mall in the US in the 1876, said joking. Since then, marketers and politicians thought a lot about ways and means to sell better their products and ideas. Nowadays, focus groups are broadly employed among marketers and advertisers. In such a scenario, brain imaging techniques, applied to human decision-making mechanisms, could be adopted to corroborate the results obtained by traditional techniques. Basically, skeptics in the neuroscience and marketing fields express the thesis that economical models and the neuroscientific techniques of brain imaging lie on so distant levels of behavioral analysis. Hence, they will hardly help themselves, reciprocally, in the understanding of problems characterizing the social and economic behavior of human beings. Although many experts among them doubt that the brain imaging techniques could be properly used for this purpose, the discipline called neuromarketing aroused great interest, as much as suspect, by means of a series of articles published on important American journals, such as *Forbes*, *The New York Times*, and *The Financial Times*. Such as newspapers gave importance to news (although often confusing and alarmist) related to the possible application of brain imaging technologies for the evaluation of the efficacy of commercial communications.

G. Vecchiato et al.: *Neuroelectrical Brain Imaging Tools*, BIOSYSROB 3, pp. 1–7.
DOI: 10.1007/978-3-642-38064-8_1

It is understandable that the idea to evaluate the neurological correlates of consumer behavior by means of the brain imaging techniques can cause a considerable arousal in marketing environments. It is important to notice that to define neuromarketing as the application of neuroimaging techniques for the analysis of consumer behavior after the exposition to advertising messages is very reductive. More properly, the neuromarketing can be defined as the field of study that applies the methodologies of the neuroscience to analyze and understand the human behavior related to market and economic exchanges. Hence, the contribution of neuroscientific methods becomes significant for the knowledge of the human behavior in the marketing scope. In fact, the main issue is to overcome the dependence of the measures nowadays used on subjects. These measures depend on the good-faith and accuracy of the experimental subject reporting his own sensations to the experimenter. Instead, the use of the brain imaging technique can distinguish the subject's cognitive experience (verbally expressed during the interview) from the activations of cerebral areas related to different, and unconscious, mental states. Experimental evidences seem to suggest that the use of the brain imaging can, in a near future, be placed side by side to classical tests today largely used in the marketing sciences.

The idea behind the present book is to illustrate the potential of both canonical and high resolution electroencephalography (EEG) techniques when applied to the analysis of brain activity related to the observation of TV commercials. In particular, we want to describe how, by using appropriate statistical analysis, it can be possible to recover significant information about the engaged scalp and cortical areas, along with variations of the activity of the autonomous nervous system, during the observation of particular key frames of the analyzed TV commercials. In the last two years we have conducted and published several studies showing how it is possible to detect hidden signs of cognitive processes and emotional involvement, such as the level of memorization and attention along with the perceived pleasantness, while watching an advertisement, by using neurophysiological recordings.

In particular, we are interested in analyzing the brain activity occurring during the "naturalistic" observation of commercial ads intermingled in a random order within a documentary. To measure both the brain activity and the emotional involvement of the investigated subjects, we conducted several experiments in order to use simultaneous EEG, galvanic skin response (GSR) and heart rate (HR) measurements. We link significant variation of these variables with the memorization, attention and pleasantness of the presented stimuli, as successively resulted from the subject's verbal interview. In order to do that, different indexes are employed to summarize the performed cerebral and autonomic measurements, later used in the statistical analysis.

The aim was to recreate, as much as possible, a "naturalistic" approach to the task in which the observer is viewing the TV screen without any particular goal in mind. In fact, the subjects were neither instructed on the aim of the task, nor aware that an interview about the observed TV commercials would be generated at the end of the movie.

Chapter 1 presents the basics of the used methodology in our neuromarketing experiments: from the basics of the cortical sources estimation to the presentation of the experimental design. Chapter 2 illustrates the scenario of a series of neuromarketing experiments we performed in the last years to investigate the EEG cerebral activations correlated with memorization, attention and emotional processing of individuals while watching TV commercials. A large space to the basics knowledge of the high resolution EEG technology is provided. Chapter 3 shows how, by using the hr EEG technique, it is possible to distinguish and describe cortical spectral activations of subjects who remembered and liked TV commercials against those of who did not. Measures and analysis of autonomic nervous system variables are also presented. Chapter 4 illustrate the details of an experiment which helped in the understanding the cortical correlates of cognitive and emotional processes during the observation of TV commercials. Chapter 5 has the aim to investigate the modulation of the Power Spectral Density (PSD) of the EEG rhythms elicited in the frontal and the prefrontal cortices during the observation of commercial advertisements. In particular, the aim is to analyse the EEG frontal asymmetrical activations occurring while subjects are watching emotional scenes of a TV advertisement. Chapter 6 illustrates the potential of the high resolution EEG techniques when applied to the analysis of brain activity related to the observation of TV commercials and Public Service Announcements (PSAs) to localize cerebral areas mostly involved. Cultural similarities and differences in advertising are inspected in Chapter 7 by studying the EEG correlates of memorization and pleasantness of Western and Eastern consumers. Chapter 8 highlights the added value of the electrical neuroimaging to improve the quality of the advertising messages. Chapter 9 reports the applications of the theoretical information previously offered in the other Chapters to the analysis of a real broadcasted TV commercials in Italy. The book offers in the Chapter 10 a description of the possible methodological advancements to be applied in a next future for the evaluation of marketing related stimuli. Computational limitations will be likely to be surpassed in the next years to make available to the modeling of these phenomena the estimation of cortical connectivity in real time.

1.1 Neuronal Responses to TV Commercials

Researchers have recently begun to investigate the neuronal mechanisms underlying a more specific form of economically relevant behavior, i.e. consumer choices concerning the selection and consumption of certain products. In this approach, one is interested in the relationship between marketing stimuli, most notably TV commercials, to which individuals are exposed. From an economic and/or marketing perspective, the aim is a better understanding of how mass consumer advertising of (established) brands affects brain systems. From a neuroscience perspective, the broad goal is a better understanding of both the neural mechanisms underlying the impact of affect and cognition on memory and the neural correlates of choice and decision-making.

Ioannides et al. (2000) have employed MEG to study the neuronal responses in subjects viewing the same TV advertisements as used by Ambler and Burne (1999). Those MEG data suggest that cognitive advertisements activate predominately posterior parietal and superior prefrontal cortices, whereas affective material modulates activity in orbitofrontal cortices, the amygdala and the brainstem. The results seem to imply that cognitive rather than affective advertisements activate cortical centers associated with the executive control of working memory and maintenance of higher-order representations of complex visual material. Interestingly, neuronal responses to affective visual material seem to exhibit a greater inter-subject variability than responses to cognitive material.

Young (2002) has used the EEG to detect putative 'branding moments' within TV commercials. These moments comprise rather short periods within the advertisement, but are assumed to do much of the 'work' in actuating advertising performance measures. Young derives a rudimentary measure of (mental) engagement from fundamental alpha, beta, and gamma rhythms present in the EEG. He has found a high correlation between moments identified by brain waves and moments identified using a behavioural, attention-sensitive method of picture sorting. This may suggest that there are indeed moments of 'special' importance within a given TV commercial.

Using EEG, Silberstein et al. (2000) and Rossiter et al. (2001) have developed a technique to measure memory encoding of visual scenes presented in TV advertisements. The results obtained in those studies suggest that visual scenes (typically > 1.5 s) that elicit the fastest brain activation in left frontal cortices are also better recognized. Those findings bear on theories of the transfer of visual information from short-term to long-term memory. In addition, those studies reinforce the notion that certain scenes within an advertisement are special in some sense.

1.2 Neuromarketing and Society

Despite many common beliefs about the inherently evil nature of marketing, the main objective of marketing is to help match products with people. Marketing serves the dual goals of guiding the design and presentation of products such that they are more compatible with consumer preferences and facilitating the choice process for the consumer. Marketers achieve these goals by providing product designers with information about what consumers value and want before a product is created. After a product emerges on the marketplace, marketers attempt to maximize sales by guiding the menu of offerings, choices, pricing, advertising and promotions. In their attempts to provide these types of inputs, marketers use a range of market research techniques, from focus groups and individual surveys to actual market tests.

In general, the simpler approaches (focus groups and surveys) are easy and cheap to implement but they provide data that can include biases, and are therefore seen as not very accurate (Beckwith and Lehmann, 1975; Day, 1975; Griffin and Hauser, 1993; Green and Srinivasan, 1990). The approaches that are more complex and therefore harder to implement, such as market tests, provide more accurate data but incur a higher cost, and the product, production and distribution systems have to be in place for market tests to be conducted.

The incorporation of neuroimaging into the decision-making sciences — for example, neuroeconomics — has spread to the realm of marketing. As a result, there are high hopes that neuroimaging technology could solve some of the problems that marketers face. A prominent hope is that neuroimaging will both streamline marketing processes and save money. Another hope is that neuroimaging will reveal information about consumer preferences that is unobtainable through conventional methods (Ariely and Berns, 2010).

1.3 Why Use Neuroelectrical Brain Imaging for Marketing?

Marketers are excited about neuroelectrical brain imaging for two main reasons. First, marketers hope that neuroimaging will provide a more efficient trade-off between costs and benefits. This hope is based on the assumptions that people are not able, or do not want, to fully articulate their preferences when asked to express them explicitly, and that consumers' brains contain hidden information about their true preferences. Such hidden information could, in theory, be used to influence their buying behavior, so that the cost of performing neuroimaging studies would be outweighed by the benefit of improved product design and increased sales. In theory, at least, brain imaging could illuminate not only what people like, but also what they will buy. Thus far, this approach to neuromarketing has focused on this post-design application, in particular on measuring the effectiveness of advertising campaigns.

The second reason why marketers are excited about neuroelectrical brain imaging is that they hope it will provide an accurate marketing research method that can be implemented even before a product exists (Figure 1).

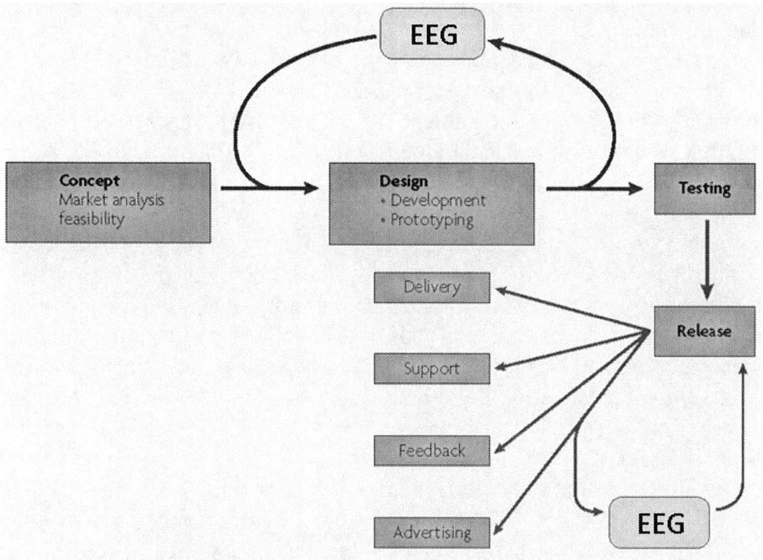

Fig. 1 Product development cycle. Neuromarketing applications of electroencephalography (EEG) can potentially enter into the product development cycle in two places. In the first, EEG can be used as part of the design process itself. Here, neural responses could be used to refine the product before it is released. In the second, EEG can be used after the product is fully designed, typically to measure neural responses as part of an advertising campaign to increase sales (adapted from Ariely and Berns, 2010).

The assumption is that neuroimaging data would give a more accurate indication of the underlying preferences than data from standard market research studies and would remain insensitive to the types of biases that are often a hallmark of subjective approaches to valuations. If this is indeed the case, product concepts could be tested rapidly, and those that are not promising eliminated early in the process. This would allow more efficient allocation of resources to develop only promising products.

1.4 The Rationale behind Employing Neuroelectrical Signals in Marketing

In these last year we assisted to an increased interest in the use of brain imaging techniques, based on hemodynamic or electromagnetic recordings, for the analysis of brain responses to the commercial advertisements or for the investigation of the purchasing attitudes of the subjects (Ioannides et al., 2000; Knutson et al., 2007, Astolfi et al., 2008a; Morris et al., 2009). The interest is justified by the possibility to correlate the particular observed brain activations with the characteristics of the proposed commercial stimuli, in order to derive conclusions about the adequacy of such ad stimuli to be interesting, or emotionally engaging, for the subjects. Standard marketing techniques so far employed involved the use of an interview and the compilation of a questionnaire for the subjects after the exposition to novel commercial ads before the massive launch of the ad itself (ad pre-test). However, it is now recognized that often the verbal advertising pre-testing is flawed by the respondents' cognitive processes activated during the interview, being the implicit memory and subject's feelings often inaccessible to the interviewer that uses traditional techniques (Zaltman, 2003). In addition, it was also suggested that the interviewer on this typical pre-testing interviews has a great influence on what respondent recalls and on the subjective experiencing of it (McDonald, 2003; Franzen and Bouwman, 2001). Taking all these considerations in mind, researchers have attempted to investigate the signs of the brain activity correlated with an increase of attention, memory or emotional involvement during the observation of such commercial ads. Researchers within the consumer neuroscience community promote the view that findings and methods from neuroscience complement and illuminate existing knowledge in consumer research in order to better understand consumer behavior (e.g., Ambler et al. 2000; Klucharev et al. 2008).

It is very well known that the hemodynamic measurements of the brain activity allow a level of localization of the activated brain structures on the order of cubic mm, being capable to detect activations also in deep brain structures such as amigdala, nucleus accumbens etc. However, the lack of time resolution, due to the delay of the cerebral blood flow's increment after the exposition to the stimuli, make the functional Magnetic Resonance Imaging (fMRI) unsuitable to follow the brain dynamics on the base of its milliseconds activity. Nevertheless, it has the problem that the recorded EEG signals are mainly due to the activity generated on the cortical structures of the brain. In fact, the electromagnetic activity elicited by deep structures advocated for the generation of emotional processing in humans is almost impossible to gather from usual superficial EEG electrodes (Nunez, 1995; Urbano et al., 1998). However, nowadays high resolution EEG technology has been developed to enhance the poor spatial information content of the EEG activity in order to detect the brain activity with a spatial resolution of a squared centimeter and the unsurpassed time resolution of milliseconds (Nunez, 1995; Bai et al., 2007; He et al., 1999; Dale et al., 2000; Babiloni et al., 2005).

It has underlined how a positive or negative emotional processing of the commercial ad is an important factor for the formation of stable memory traces (Kato et al., 2009). Hence, it became relevant to infer the emotional engage of the subject by using indirect signs for it. In fact, indirect signs of emotional processing could be gathered by picking variations of the activity of the anatomical structures linked to the emotional processing activity in humans, such as the activity of sweat glands on the hands and/or the variation of the heart rate (Baumgartner et al., 2006), along with the tracking of the variations of the pre- and frontal cortex (PFC and FC respectively; Davidson and Irwin, 1999).

In particular, by monitoring autonomic activity using devices able to record the variation of the skin conductivity and the heart rate it is possible to assess the "internal" emotional state of the subject. In fact, the GSR is actually viewed as a sensitive and convenient measure of indexing changes in sympathetic arousal associated with emotion, cognition, and attention (Critchley, 2002). Studies using functional imaging techniques (Critchley, 2002; Nagai et al., 2004) have related the generation and the maintenance of the electrodermal activity level to specific brain areas. These specific regions are the ventromedial prefrontal cortex, orbitofrontal cortex, left primary motor cortex, and the anterior and posterior cingulate, which have been shown to be associated with emotional and motivational behaviors (Critchley, 2002; Nagai et al., 2004). Such findings indicate the close association of peripheral and central measures of arousal, re-emphasize the close connections between electrodermal activity, arousal, attention, cognition and emotion. In addition, the link between the heart rate or the Heart Rate Variability (HRV) and the sympatho/vagal balance has been already suggested (Malik et al., 1996; Malliani, 2005; Montano et al., 2009).

Finally, the PFC region is structurally and functionally heterogeneous but its role in emotion is well recognized (Davidson, 2002). EEG spectral power analyses indicate that the anterior cerebral hemispheres are differentially lateralized for approach and withdrawal motivational tendencies and emotions. Specifically, findings suggest that the left PFC is an important brain area in a widespread circuit that mediates appetitive approach, while the right PFC appears to form a major component of a neural circuit that instantiates defensive withdrawal (Davidson, 2004; 2000).

Chapter 2
Methodology of a Typical "Neuromarketing" Experiment

The present Chapter illustrates the scenario of a series of neuromarketing experiments we performed in the last years to investigate the EEG cerebral activations correlated with memorization, attention and emotional processing of individuals while watching TV commercials. A large space to the basics knowledge of the high resolution EEG technology is provided. Specifically, we show how it is possible to enhance the spatial resolution of the EEG by solving the related linear inverse problem in order to obtain an estimation of the cortical sources. In addition, the theory we used to perform the spectral statistical analysis is also presented along with a direct example on fake data.

2.1 An Overview on the High Resolution EEG Technique

Information about the brain activity can be obtained by measuring different physical variables arising from brain processes, such as the increase in consumption of oxygen by the neural tissues or a variation of the electric potential over the scalp surface. All these variables are connected in direct or indirect way to the neural ongoing processes, and each variable has its own spatial and temporal resolution. Hence, the different neuroimaging techniques are then constrained by their spatio-temporal resolution, depending on the monitored variables. Human neocortical processes involve temporal and spatial scales spanning several orders of magnitude: from the rapidly shifting somatosensory processes characterized by a temporal scale of milliseconds and a spatial scale of few square millimeters, to the memory processes, involving time periods of seconds and spatial scale of square centimeters. Today, no neuroimaging method provide a spatial resolution on a millimeters scale and a temporal resolution on a millisecond scale at the same time.

Electroencephalography is an interesting technique that presents a high temporal resolution, on the millisecond scale, adequate to follow a particular kind of brain activity (to deepen this topic, far from the goal of the present book, we suggest Michel and Murray, 2012; Hallez et al., 2007; Michel et al., 2004; for a review). Unlikely, this technique has a relatively modest spatial resolution, beyond the centimeter, because of the inter-sensor distances and the fundamental laws regulating the electromagnetism (Nunez, 1995).

G. Vecchiato et al.: *Neuroelectrical Brain Imaging Tools*, BIOSYSROB 3, pp. 9–31.
DOI: 10.1007/978-3-642-38064-8_2

The simultaneous activation of an entire population of neurons can generate an electric signal detectable on the head surface with electrodes placed on the scalp.

In order to estimate the cortical activity, the EEG signal has to be measured in different scalp sites; the most common measurement system is the international montage 10-20.

Generally, the standard EEG analysis, using 20-30 electrodes, brings to a spatial resolution of about 6÷7 cm.

High resolution EEG is a technology used to increase the spatial resolution of the EEG potentials recorded on the scalp (Le and Gevins, 1993; Gevins et al., 1994; Nunez, 1995): in this case, the data are acquired using 64-128 electrodes and then processed to remove the effects of the attenuation caused by the low conductivity characterizing the head structures .

2.1.1 The Head and the Source Model

A key point of the highresolution EEG technologies is the availability of the accurate model of the head as a volume conductor by using anatomical MRI. These images are obtained by using the MRI facilities worldwide available in research and clinical institutions. Reference landmarks such as nasion, inion, vertex, and preauricular points may be labeled using vitamin E pills as markers. T1-weighted MR images are typically used since they present maximal contrast between the structures of interest.

Contouring algorithms allow the segmentation of the principal tissues (scalp, skull, dura mater) from the MR images (Dale et al., 1999; Babiloni et al., 1997, 2000). Separate surfaces of scalp, skull, dura mater and cortical envelopes are extracted for each experimental subject, yielding a closed triangulated mesh. This procedure produces an initial description of the anatomical structure by using several hundred thousand points –too computationally expensive for the subsequent mathematical processing. These structures are thus down-sampled and triangulated to produce scalp, skull and dura mater geometrical models with about 1000-1300 triangles for each surface. These triangulations were found to be adequate to model the spatial structures of these head tissues.

A different number of triangles are used in the cortical surface modeling, since its envelope is more convoluted than the scalp, skull and dura mater structures. In order to properly shape the cerebral cortex, the number of triangles used to model such a surface varies between 5000 and 6000 . In order to allow co-registration with other geometrical information, the coordinates of the triangulated structures are referred to an orthogonal coordinate system (x, y, z), which is based on the positions of nasion and pre-auricular points, previously extracted from the MR images. For instance, the midpoint of the line connecting the pre-auricular points can be set as the origin of the coordinate system and, the y axis going through the right pre-auricular point, the x axis lying on the plane determined by nasion and pre-auricular points (pointing to the nose) and the z axis is normal to this plane (pointing up). Once the model of scalp surface has been generated, the integration of the electrodes positions is accomplished by using the information about the sensor locations produced by the 3-D digitizer. The sensors positions on the scalp model are determined by using a non-linear fitting technique.

Figure 2 shows the results of the integration between the EEG scalp electrodes position and a realistic head model.

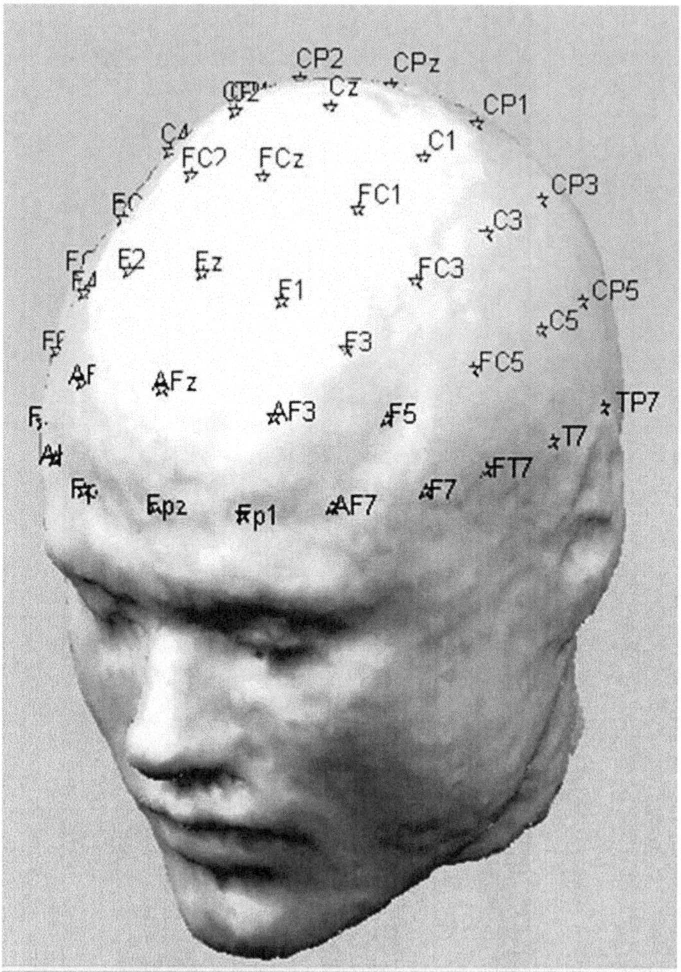

Fig. 2 Integration between EEG scalp electrodes position and a realistic head model. Realistic head model has been generated by using the T1-weighted Magnetic Resonance Images (MRIs) of the subject. In case no MRIs of the subject's head are available, it is still possible to determine the actual electrode positions used for the EEG recording with the average head model, i.e. a standard head taken as an average of the MRIs of the head of 150 subjects. Such average head model is available from the McGill University Web site.

Let us assume an EEG recording performed with a set of M electrodes disposed on the scalp surface. In the following, we indicate with b the vector of M electrical measurements recorded from the scalp. In order to perform the estimation of the cortical activity we also need a model of the head volume conductor. This model is used to approximate the propagation of the potential from the modeled neural generators to the measurement sensors. The head model can be of spherical or elliptical shape or can recall the realistic head shape. Since we have to estimate the source activity from non-invasive EEG measurements, we need a mathematical model for the neural sources. In the analysis of EEG and MEG, the largely used mathematical model for the neural source is the current dipole. Such model is largely used in literature since approximates very well the activity of relatively small patches of the cortical tissue. In this particular context, we used many current dipoles located in the brain along the entire cerebral volume or along its cortical surface. If we would like to model the entire brain volume, we had to divide it in voxels and then placing a triplet of orthogonal current unitary dipoles at each voxel position. Another source model takes into account only the cortical geometry, to constraint the dipoles to lie orthogonal to the modeled cortical surface. In the following, we deal with the problem of source estimation by using a representation of the cortical surface, and we use a set of dipoles disposed along such surface as source model (see Figure 3). However, all the mathematical formulas presented here still hold also in the case the neuronal space is divided in voxels, attempting to model also the subcortical structures.

Each dipole placed inside the volume conductor model of the head has a unitary strength and different direction, according to the local cortical geometry or to the adopted reference coordinate system. There is no limitation for the number of sources placed inside the head model, it only depends on the modeling capabilities of the used computational system. In the following, we indicate with N the number of dipoles whose strength is to be estimated from the M-dimensional measurement vector b. The typical values for N are between 1000 and 7000, while the values for M are in the range 64-256. We indicate as x the n-dimensional vector of the unknown current strengths for the dipoles.

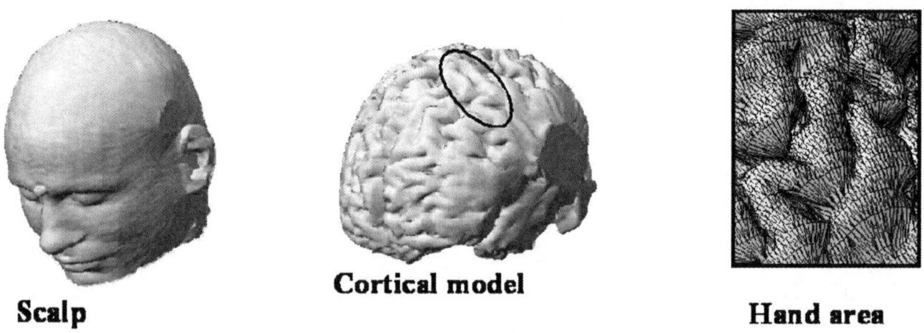

Fig. 3 Realistic head model for linear inverse source estimation. Note the particular showing the hand area tessellated with triangles (right Figure). At the center of each triangle, a dipole with unitary strength and perpendicular to the triangle was posed.

2.1.2 *Cortical Source Estimation*

In the estimation of neuronal activity from non-invasive measurements, we have to use a mathematical model for the description of the propagation of the potential distribution from each modeled sources to the sensors positions. In other words, we have to compute the potential distribution occurring on the set of the M sensors over the head model due to the i-th unitary dipole placed at the i-th cortical location. Such predictions can be made with the aid of analytical or boundary-element formulations, depending on the shape of the head model used. While the equation for the potential distribution for a three-layered spherical head model can be found in literature, here we report the equations that compute the potential value due to a dipole inside a realistic head model over a point located on the scalp surface.

Let a head model be constituted by electrically homogeneous and isotropic compartments simulating scalp, skull, and dura mater. The forward solution specifying the potential distribution (V) on these compartments S_k (k = 1,...3) due to a dipole is given by the Fredholm integral equation of the second kind:

$$(\sigma_i^- + \sigma_i^+)V(\vec{r}) = \tag{1}$$

$$= 2V_0(\vec{r}) + \frac{1}{2\pi}\sum_{j=1}^{m}(\sigma_i^- + \sigma_i^+)\int_{S_j} V(\vec{r}\,')\,d\Omega_{\vec{r}}(\vec{r}\,')$$

with

$$d\Omega_{\vec{r}}(\vec{r}\,') = \frac{\vec{r}\,'-\vec{r}}{|\vec{r}\,'-\vec{r}|^3}\,d\vec{S}_j(\vec{r}\,') \tag{2}$$

where (i) $V_0(\vec{r})$ is the potential due to a dipole located in an infinite homogeneous medium; (ii) σ_i^- is the conductivity inside the surface S_j of the multi-compartment head model; (iii) σ_i^+ is the conductivity outside the surface S_j; (iv) m is the total number of compartments within the head model; and (v) $d\Omega_{\vec{r}}(\vec{r}\,')$ is the solid angle subtended by the surface element dS located in $\vec{r}$ (point of observation $\vec{r}\,'$). A numerical solution of the Fredholm integral equation can be obtained by decomposing the surfaces S_k (k = 1,...3) into triangle panels and by using boundary element techniques. With the boundary-element techniques, a discrete version of the Fredholm integral equation is given by:

$$\mathbf{v} = \mathbf{g} + \mathbf{\Omega v} \tag{3}$$

where the elements of matrix $\mathbf{\Omega}$, vector **v**, and vector **g** are defined as follows: (i) **v**i is the potential value in the center of mass of the i-th triangle; (ii) **g**i is the potential value generated by a source in the center of mass of the i-th triangle; and (iii) $\mathbf{\Omega}$ij is the matrix element proportional to the solid angle subtended by the j-th triangle at the center of mass of the i-th triangle. The numerical solution of the Fredholm integral equation can be improved using the deflation procedure. The linear system of equation 3 is singular since the potential distribution generated on the scalp compartment by an equivalent dipole is determined up to a constant. This singularity can be removed by using a deflation procedure that yields the potential distribution (V) on the compartment surfaces S_k (k = 1,...3).

In the following, we indicate as **A**i the potential distribution over the M sensors due to the unitary i-th cortical dipole. The collection of all the M-dimensional vectors **Ai**, (i = 1,..., N) describes how each dipole generates the potential distribution over the head model. This collection is called the lead field matrix **A**.

With the definitions provided above, we can say that the estimate of the strength of the modeled dipolar source strength **x** from the non-invasive set of measurement **b**, is obtained by solving the following linear system at a particular instant in time t:

$$\mathbf{Ax} = \mathbf{b} + \mathbf{n} \tag{4}$$

The Eq. 4 provides an estimation of the dipole source configuration **x** at time t that generates the measured EEG potential distribution **b** in the same instant. The system also includes the measurement noise **n**, assumed to be normally distributed (Nunez, 1995; Grave de Peralta Menendez and Gonzales Andino 1998). **A** is the lead field matrix, where each j-th column describes the potential distribution generated on the scalp electrodes by the j-th unitary dipole. **A** matrix has the dimension of the number of electrodes employed times the number of cortical sources used in the model. The vectors **x** has the dimension of the number of cortical sources employed in the estimation (typical values between 3,000 to 5,000 sources), while the vectors **b** and **n** have both the dimension of the number of electrodes, M, employed for the recordings.

This is a strongly underdetermined linear system, in which the number of the unknown variables (i.e. the dimension of the vector **x**), is greater than the number of measurement **b** of about one order of magnitude. In this case, from the linear algebra we note that there are infinite solutions for the vector of dipole strengths **x**. All these solutions explain in the same way the data vector **b**. Furthermore, the linear system is ill conditioned as results of the substantial equivalence of the several columns of the electromagnetic lead field matrix **A**. In fact, we know that each column of the lead field matrix arise from the potential distribution generated by the dipolar sources that are located in similar position and orientation along the used cortical model. Regularizing the inverse problem consists in attenuating the oscillatory modes generated by vectors associated with the smallest singular values of the lead field matrix **A**, introducing supplementary and a priori information on the sources to be estimated. In the following, we characterize with the term "solution space" the space in which the "best" current strength solution **x** will be found. The "measurement space" is the vectorial space in which the vector **b** of the gathered data is considered. The solution of the linear problem defined in the Eq. 4 with the variation approach is based on the idea of selecting metrics from the solution space and the measurement space, respectively.

Before proceeding to the derivation of a possible solution for the present problem, we recall a few useful algebra definitions to better understand the following notations. A more complete introduction to the theory of vector spaces is out of the scope of this chapter, and the interested readers could refer to related textbooks (Spiegel, 1978; Rao and Mitra 1977). In a vector space provided with a definition of an inner product (•,•), it is possible to associate a value or modulus to a vector **b** by using the notation:

$$(\mathbf{b}, \mathbf{b}) = \|\mathbf{b}\| \tag{5}$$

Any symmetric positive definite matrix **M** is said a metric for the vector space furnished with the inner product (•,•) and the squared modulus of a vector **b** in a space equipped with the norm **M** is described by

$$||\mathbf{b}||_{\mathbf{M}}^{2} = \mathbf{b}^{\mathrm{T}}\mathbf{M}\mathbf{b} \tag{6}$$

With these recalls in mind, we now face the problem to derive a general solution of the problem previously described under the assumption of the existence of two distinct metrics **N** and **M** for the source and the data space, respectively. These two metrics are characterized by symmetric matrices and express our idea of closeness in the same spaces. With this approach, the minimization function is composed of two terms, one that evaluates how well the solution explains the data, and the other that measures the closeness of the solution to an a priori selected. Since the system is undetermined, infinite solutions exist. However, we are looking for a particular vector solution **x** that has the following properties: 1) it has the minimum residual in fitting the data vector **b** under the norm M in the data space 2) it has the minimum strength in the source space under the norm N. To take into account these properties, we have to solve the problem utilizing the Lagrange multiplier and minimizing the following functional that express the desired properties for the sources **x** (Dale and Sereno, 1993; Tichonov and Arsenin 1977; Menke, 1989; Grave de Peralta Menendez and Gonzales Andino, 1998; Liu, 2000):

$$\phi = ||\mathbf{A}\mathbf{x} - \mathbf{b}||_{\mathbf{M}}^{2} + \lambda^{2}||\mathbf{x}||_{\mathbf{N}}^{2} \tag{7}$$

where M, N are the matrices associated with the metrics of the data and of the source space, respectively, λ is the regularization parameter and $||\mathbf{x}||_{\mathbf{M}}$ represents the M norm of the vector **x**. Thus, **M** is a square matrix whose dimensions are the number of sensors, while **N** is a square matrix whose dimensions are the number of cortical sources employed. The solution of the problem depends on adequacy of the data and source space metrics. Under the hypothesis of M and N positive definite, the solution is given by taking the derivatives of the functional ϕ and setting it to zero. After few straightforward computations the solution is:

$$\hat{\mathbf{x}} = \mathbf{G}\mathbf{b} \tag{8}$$

$$\mathbf{G} = \mathbf{N}^{-1}\mathbf{A}'(\mathbf{A}\mathbf{N}^{-1}\mathbf{A}' + \lambda\mathbf{M}^{-1})^{-1} \tag{9}$$

where **G** is called the pseudo-inverse matrix, or the inverse operator, that maps the measured data **b** onto the source space. By construction, the pseudo-inverse operator **G** has the dimensions of the number of cortical sources times the number of electrodes employed. An optimal regularization of this linear system was obtained by the L-curve approach [He et al, 2006].

Note that the requirements of positive definite matrices for the metric **N** and **M** allow to consider their inverses. Last equation stated that the inverse operator **G** depends on the matrices **M** and **N** that describe the norm of the measurements and the source space, respectively. The metric **M**, characterizing the idea of closeness in the data space, can be particularized by taking into account the sensors noise level by using the Mahalanobis distance (Grave de Peralta Menendez and Gonzales Andino, 1998). If no a priori information is available for the solution of linear inverse problem, the matrices **M** and **N** are set to the identity, and the minimum norm estimation is obtained (Hämäläinen and Ilmoniemi, 1984). However, it was recognized that in this particular application the solutions obtained with the minimum norm constraints are biased toward those sources that are located nearest to the sensors. In fact, there is a dependence of the distance on the law of potential (and magnetic field) generation and this dependence tends to increase the activity of the more superficial sources while depresses the activity of the sources far from the sensors. The solution to this bias was obtained by taking into account a compensation factor for each dipole that equalizes the "visibility" of the dipole from the sensors. Such technique, called column norm normalization by Lawson and Hanson (1974), was used in the linear inverse problem by Pascual-Marqui (1995, 2002) and then adopted largely by the scientists in this field. With the column norm normalization the inverse of the resulting source metric is

$$(\mathbf{N}^{-1})_{ii} = \|\mathbf{A}_i\|^{-2} \tag{10}$$

where $(\mathbf{N}^{-1})_{ii}$ is the i-th element of the inverse of the diagonal matrix **N** and all the other matrix elements $\mathbf{N}_{ij}$ are set to 0. The L2 norm of the i-th column of the lead field matrix **A** is denoted by $\|\mathbf{A}_i\|$. In this way, dipoles close to the sensors, and hence with a large $\|\mathbf{A}_i\|$, will be depressed in the solution of the inverse problem, since their activations are not convenient from the point of view of the functional cost. The use of this definition of matrix N in the source estimation is known as weighted minimum norm solution (Grave de Peralta et al., 1997; Pascal-Marqui, 1995).

2.1.3 Cortical Power Spectrum Computation

Here we propose an alternative procedure, that is based on the estimation of spectral power of cortical signals based on spectral power of scalp measurements.

Let:

$$\mathbf{b}(t) = \mathbf{A}\mathbf{x}(t), \tag{11}$$

where $t \in [\tau, \tau + \Delta\tau]$, be the vector of scalp measurements in a given time window, with $\Delta\tau$ being the sampling time, and

$$\mathbf{b}(f) = \mathbf{A}\mathbf{x}(f) \tag{12}$$

the corresponding vector of cortical estimates, if $\mathbf{b}(f)$ is the Fourier transform of $\mathbf{b}(t)$.

We also have

$$\hat{\mathbf{x}}(f) = \mathbf{G}\mathbf{b}(f) \tag{13}$$

where **G** is the pseudo inverse of the lead field matrix **A** and $\hat{\mathbf{x}}(t)$ the corresponding vector of cortical estimates at time t given by Eq.8. We define the matrix of Cross-Power Spectral Densities (**CSD**) as the matrix whose element (i, j) is the cross-spectrum of i-th and j-th channel of the signal. By using the exponent 'sens' to indicate the sensors' measurements and the exponent 'src' to indicate the cortical sources we have:

$$\mathbf{CSD}^{(sens)}(f,\tau) = \mathbf{b}(f,\tau)\mathbf{b}^{\mathrm{H}}(f,\tau) \tag{14}$$

where the $\mathbf{b}^{\mathrm{H}}(f,\tau)$ is the conjugate transposed (Hermitian) of $\mathbf{b}(f,\tau)$. Analogously:

$$\mathbf{CSD}^{(\mathrm{src})}(f,\tau) = \hat{\mathbf{x}}(f,\tau)\hat{\mathbf{x}}^{\mathrm{H}}(f,\tau) \tag{15}$$

By using Eq.13 we could use the pseudo-inverse **G** to obtain:

$$\mathbf{CSD}^{(src)}(f,\tau) = \mathbf{Gb}(f,\tau)\mathbf{b}^{\mathrm{H}}(f,\tau)\mathbf{G}' = \mathbf{G}\cdot\mathbf{CSD}^{(src)}(f,\tau)\cdot\mathbf{G}' \tag{16}$$

If $\mathbf{b}(t)$ and $\hat{\mathbf{x}}(t)$ are not deterministic signals, but rather we have several trials (realizations) of a stochastic process, Eq 16 holds if we substitute **CSDs** with their expected values or with their estimates (i.e. $\left\langle \mathbf{CSD}^{(\mathrm{src})}(f,\tau)\right\rangle$).

In case we were interested in calculating the (Auto-) Power Spectral Densities (PSDs) of estimated cortical sources, we only need to compute the diagonal of $\mathbf{CSD}^{(\mathrm{src})}(f,\tau)$:

$$\mathbf{PSD}_j^{(src)}(f,\tau) = \mathbf{G}_j\cdot\mathbf{CSD}^{(sens)}(f,\tau)\cdot\mathbf{G}_j{}' \tag{17}$$

where the variable j indicates the number of sources.

The spectral resolution of this method is inversely proportional to $\Delta\tau$.

The level of noise in the EEG linear inverse solutions can be addressed by estimation of the 'projection' $\mathbf{n}^{(src)}(t)$ of the EEG noise $\mathbf{n}^{(sens)}(t)$ onto the cortical surface by means of the computed pseudo-inverse operator **G** (as described in Eq. 8); the variance of the noise on the estimated source strength $\hat{x}_j(t)$ is given by:

$$\sigma_j^{2(src)} = C_{jj}^{(src)} = \langle n_j^{(src)}, n_j^{(src)}\rangle = \mathbf{G}_j\cdot\mathbf{C}^{(sens)}\cdot\mathbf{G}_j' \tag{18}$$

where $\mathbf{G}_j$ is the j-th row of the pseudo-inverse matrix, $\mathbf{C}^{(sens)}$ is the EEG noise covariance matrix ($C_{ij}^{(sens)} = \left\langle n_i^{(sens)},\ n_j^{(sens)}\right\rangle$), and $\langle\cdot,\ \cdot\rangle$ is the expectation operator. A common choice for the estimation of the sensor noise covariance matrix is to select an interval of data (baseline) where no task-related activity is supposed to occur and thus all signals are believed to be noise:

$$\mathbf{n}^{(sens)} \equiv \mathbf{b}(\mathrm{t}), \quad t \in baseline \tag{19}$$

According to the DSPM approach (Dale et al 2000), the following normally-distributed cortical z-score estimator can be obtained for each j-th cortical location and for each time point considered:

$$z_j(t) = \frac{\hat{x}_j(t)}{\sigma_j^{(src)}} = \frac{\mathbf{G}_j \cdot \mathbf{b}(t)}{\sqrt{\mathbf{G}_j \cdot \mathbf{C}^{(sens)} \cdot \mathbf{G}_j'}} \tag{20}$$

This allows us to assess quantitatively the ratio between the estimated cortical activity $\hat{\mathbf{x}}$ at a particular instant in time and the amount of noise at cortical level, quantified through the standard deviation of its estimate. Values of z exceeding a given threshold represent levels of estimated cortical activity that are unlikely owing to chance alone but are related to the task performed by the experimental subject. For instance, the threshold for the z-score level at 5% is $z_{5\%} = 1.96$.

In this particular application, we considered as baseline the estimated cortical activity during the viewing of the documentary. Here, we extend the DSPM approach to the analysis of the power spectra variations during the experimental task. The computation of the z-score level in the spectral case is performed according to the following:

$$\Sigma_j^{2(src)}(f) = Var\left\{PSD_j^{(src)}(f,\tau)\right\}_{\tau \in baseline}$$

$$\mathrm{M}_j^{(src)}(f) = Mean\left\{PSD_j^{(src)}(f,\tau)\right\}_{\tau \in baseline}$$

$$Z_j(f,\tau) = \frac{PSD_j^{(src)}(f,\tau) - M_j^{(src)}(f)}{\Sigma_j^{(src)}(f)} = \frac{\mathbf{G}_j \cdot \mathbf{CSD}^{(sens)}(f,\tau) \cdot \mathbf{G}_j{}' - M_j^{(src)}(f)}{\sqrt{\Sigma_j^{2(src)}(f)}} \tag{21}$$

where $Var\left\{PSD_j^{(src)}(f,\tau)\right\}_{\tau \in baseline}$ indicates the variance of the estimate of the spectral density of the EEG measurements during the baseline period at the considered frequency f, and $Z_j(f,\tau)$ is the z-score for the j-th current dipole at frequency f, while the inverse operator **G** is the same used for the temporal case. The $Z_j(f,\tau)$ is a z-score variable for construction.

Following these calculations it has been possible to obtain and analyse spectral Z variables for the canonical frequency bands of interests: Theta (4-7 Hz), Alpha (8-12), Beta (13-24), Gamma (25-40). It must be noted that such z-score transformation of the cortical data is different from the z-score transformation adopted on the GSR, HR and GFP values described in the previous paragraph, since these last z-score referred to autonomic and EEG signals from the scalp surface, while the z-score transformation described by eq. 20 and 21 are related to the estimated cortical signals.

In order to proper deal with the issue of the execution of multiple univariate statistical tests, like those designed in Eq. 21, we used the Bonferroni correction for the evaluation of the significant z-scores (Zar, 2009).

2.1.4 Statistical Analysis

During the execution of an experiment by using EEG measurements, typically subjects repeat the task a variable number of times (often called trials) in order to collect enough EEG data to allow a statistical validation of the results.

Let **S** be the matrix of the power spectra of the cortical sources computed, which has the dimension equal to the number of sources times the frequency bin used times the number of trials recorded. We compute the average of the power spectral values related to i-th dipole within the j-th frequency band of interest (Theta, 3-7 Hz; Alpha, 8-12 Hz; Beta, 13-29 Hz; Gamma, 30-40 Hz); this operation is repeated for each source and frequency band. By this way, for each frequency band, we have a matrix $\bar{\mathbf{S}}_j$ (with dimension sources times trials) which represents the distribution along the number of trials of the mean spectral power of each cortical sources.

The aim of the procedure is to find the differences between the cortical power distributions related to two different experimental tasks performed by the subject, say task A and task B. For this reason, we compute the matrices $\bar{\mathbf{S}}_j$ related to the EEG data recording during the task A and during the task B, and we referred to them as $\overline{\mathbf{S}_j^A}$ and $\overline{\mathbf{S}_j^B}$. Successively, we perform a statistical contrast between such spectral matrices $\overline{\mathbf{S}_j^A}$ and $\overline{\mathbf{S}_j^B}$ by using appropriate univariate statistical tests (such as the Student's test with the correction for multiple comparisons).

Let us consider the i-th source and the j-th frequency band, we denote with $\mu_A^{i,j}$ and $\mu_B^{i,j}$ the mean values of the cortical power spectra distribution. In the following we want to verify the null hypothesis H_o that such means values are statistically similar, i.e.

$$H_0: \mu_A^{i,j} = \mu_B^{i,j} \text{ and } H_A: \mu_A^{i,j} \neq \mu_B^{i,j}$$

where H_A is the alternative hypothesis to test that such differences are instead significantly different.

Under the assumptions that i) the two samples came from normal (Gaussian) population, and ii) the populations have equal variances, the Student's t value for testing the previous hypotheses can be expressed as:

$$t = \frac{\overline{X_A} - \overline{X_B}}{s_{\overline{X_A} - \overline{X_B}}} \tag{22}$$

The quantity $\overline{X_A} - \overline{X_B}$ is simply the difference between the two means, and $S_{X_A - X_B}$ is the standard error of the difference between the samples. The last quantity is a statistic that can be calculated from the sample data and it is an estimate of the standard error of the population indicated as $\sigma_{X_A - X_B}$. It can be shown mathematically that the variance of the difference between two independent variables is equal to the sum of the variances of the two variables, so that the standard error of the population could be computed as the sum of the standard error of the two groups as follows $\sigma_{X_A - X_B} = \sigma_{X_A} + \sigma_{X_B}$.

Independence means there is no correlation between the two variables *A* and *B*. As $\sigma_X = \sqrt{\frac{\sigma^2}{n}}$, where n is the population dimension, we can write:

$$\sigma_{X_A - X_B} = \sqrt{\frac{\sigma_A^2}{n_A} + \frac{\sigma_B^2}{n_B}} \tag{23}$$

As the two-sample Student's t test requires the homoscedasticity of the variances of the two samples (i.e. we assume $\sigma_A^2 = \sigma_B^2 = \sigma^2$), we can write

$$\sigma_{X_A - X_B} = \sqrt{\frac{\sigma^2}{n_A} + \frac{\sigma^2}{n_B}} \tag{24}$$

Thus, to calculate the estimate of $\sigma_{X_A - X_B}$ an estimates of σ^2 is required. Denoting with S_A^2 and S_B^2 the statistically similar estimators of the variance σ^2, we compute the pooled variance S_P^2 to obtain the best estimate of σ^2:

$$S_P^2 = \frac{(n_A - 1)S_A^2 + (n_B - 1)S_B^2}{n_A + n_B - 2} \tag{25}$$

and

$$S_{X_A - X_B}^2 = \frac{S_P^2}{n_A} + \frac{S_P^2}{n_B} \tag{26}$$

Thus

$$S_{X_A - X_B} = \sqrt{\frac{S_P^2}{n_A} + \frac{S_P^2}{n_B}} \tag{27}$$

Finally, we obtain:

$$t = \frac{\overline{X_A} - \overline{X_B}}{\sqrt{\frac{S_P^2}{n_A} + \frac{S_P^2}{n_B}}} \tag{28}$$

The results of the previous statistical analysis can be represented on a realistic model of the cortex of the subjects, for each of the frequency band of interest. Over the cortex we will see only the statistically significant cortical activations, pointing out the differences between the characteristic activation during the considered tasks.

To now, the assumptions of the normality and homoscedasticity of the estimated spectra are assumed in order to perform the statistical analysis with the Student's test. It might be argued which test could be used in the case in which such assumptions did not hold. In this respect the good news is that the robustness of the Student's test from the assumption of homoscedasticity and normality are generally very high, and hence we can use such test although there are no precise information related to the gaussianity and the homoscedasticity of the data (Zar, 2009). However, the appropriate statistic to deal with the heteroscedastic case is the Sattertwaite or Cochran and Cox estimation of the standard error to insert in the formulation above presented. Such calculations, although lead to results very similar to those obtained with the standard Student's approach, could be find in standard statistic textbook (Zar, 2009).

2.1.5 *The Issue of Multiple Univariate Comparisons*

As last statistical issue, it is well know that the performance of many univariate statistical tests (like those presented in this application, one for each cortical dipole modeled) could easily generate the appearance of "false positive" results (known as type I error or "alpha inflaction"). This means that statistically significant differences were found although there are no real differences between the two analyzed samples just because the high number of univariate tests performed.

Strictly speaking, a statistical significance (e.g., $p<0.05$) means that we are sure that the difference detected in the sensors analyzed in the two experimental conditions are not due to chance alone 95 times over 100. However, there are 5 cases over 100 in which we are declaring that the differences we found in EEG or MEG activity are significant while they are just due to the chance. This risk, i.e. the risk to declare statistically significant what it is not significant at all, is known as Type I error in statistical literature. At first sight, this risk seemed not too much high and then acceptable in order to draw appropriate conclusions on the phenomena to be studied with the EEG/MEG data collection performed. However, if the N univariate tests performed are independent we have that the risk to declare significant a comparison by chance alone at $p = 0.05$ is equal to N times 0.05. Depending on N, the risk to declare as significant results due to the chance alone became substantial. In particular, let us suppose to generate a statistical comparison with a significance level of $p = 0.05$ between two populations during a cognitive task. Let us consider as a source space an average head model with 3,000 current dipoles displaced perpendicularly to the cortical surface. In this case, the number of elements in the source space that became significant by chance alone is equal to 0.05 * 3,000 = 150, if the multiple comparisons are executed for each one of these dipoles. This number could be also multiplied by 3 if the statistical tests are performed on the three components of each current dipole, reaching almost 500 locations in which the result of the application of the multiple univariate statistics is "inflated" by the type I error (Zar, 2009). As described above, the scientific literature is full of solutions for this problem.

The field of statistic provides different body of techniques that could be usefully applied in the context of the analysis of the neuroelectromagnetic statistical mapping. Some of these techniques are described briefly in the following.

2.1.5.1 Family-Wise Error Rate

False positives are results of statistical tests that exceed the threshold employed by chance alone. The family-wise error rate represents the probability of observing one or more false positives after carrying out multiple significance tests. Using a Family-Wise Error Rate (FWER) of 0.05 would mean that there is a 5% chance that one or more false positives could occur across the entire set of hypothesis tests. The Bonferroni adjustment is probably the most widely known FWER control since it is the adjustment method for the control of the occurrence of false positives that most investigators are familiar with. Due to the scope of this paper, although other methodologies are available for FWER (Brett et al., 2004; Nichols and Hayasaka, 2003), in the following sections we limit the discussion of this methodology to False Discovery Rate and the Bonferroni adjustment, with also its less stringent variant known as Bonferroni-Holm procedure. They are also implemented in popular software packages for the estimation of the inverse problem, such as sLORETA (Pasqual-Marqui, 2002).

2.1.5.2 False Discovery Rate

If the issue is the limitation of the occurrence of false positives, the FWER methodologies make their duty, although this happens at the expenses of the statistical power of the test (i.e., there is the need of less variance in the data to detect significant real differences). However, other methods are available in literature, but here just the technique of False Discovery Rate will be discussed. Another approach to multiple comparisons adjustment is to place limits on the False Discovery Rate (FDR). Using a false discovery rate of 0.05 would mean that at most 5% of the detected results are false positives (Benjamini and Hochberg, 1995; Benjamini and Yekutieli, 2001). This control on the number of false positives could be not optimal, especially with respect to the control performed by the FWER techniques, but it has been considered a more ideal balance between statistical power and multiple comparisons control because of its less conservative approach, at least in studies of brain activity that used the hemodynamic signals (Genovese et al. 2002).

2.1.5.3 Bonferroni Adjustment

The Bonferroni correction (Bonferroni, 1936) starts stating that if we perform N univariate tests, each one of them with an unknown significant probability β, the probability P that at least one of the test is significant is given by:

$$p < N\beta \tag{29}$$

due to the Boole's inequality (Zar, 2009). In other words this means that if N = 20 tests are performed with the usual probability β = 0.05, at least one of them became significant statistically by chance alone.

However, the Bonferroni adjustment required that the probability P for which this event could occur (i.e. one result will be statistically significant by chance alone) could equal to α, with a tipical value of 0.05. By using the equation (29), the single test will be performed at a probability β = α / N. In the case of N = 20 and α = 0.05 we have β* = 0.05 / 20 = 0.0025. This β* is the actual probability at which the statistical tests are performed in order to conclude that all of the tests are performed at a 0,05 level of statistical significance, Bonferroni corrected for multiple comparisons.

The Bonferroni adjustment is quite flexible since it does not require the hypothesis of independence of the data to be applied. However, there is some consensus that Bonferroni may be too conservative both for EEG and MEG data as well as for the analysis of hemodynamic data (Logan, Geliazkova, & Rowe, 2008). In fact, the values obtained at each sensor, as well as in the source space, often are not really statistically independent one to each other (Nunez, 1995; Grave de Peralta and Gonzalez Andino, 1999). For instance, in the neuroelectromagnetic modality it has been demonstrated that the lead field matrix (i.e. the matrix describing all the potential distributions occurring at the scalp level for the unitary current dipoles employed in the source space) is rank-deficient, meaning that the activity propagated by nearby dipoles at the cerebral level is equivalent on the scalp sensors. This last effect is commonly known as volume conduction effect. The same dependence between the cerebral activity occurring in nearby voxel has been also advocated by using the hemodynamic measurements (Logan, Geliazkova, & Rowe, 2008). This correlation causes the corrected Bonferroni threshold to be unnecessarily high, potentially eliminating valid results, "removing the baby with the dirty water".

Statisticians provide an adjustment for this non necessary severity, the so called Bonferroni-Holm adjustment. It must be noted that recently an adjustment of the nominal level of significance α for the distributed estimation of brain sources has been proposed by dividing it according to the number of independent sensors employed (Grave de Peralta et al., 2004).

2.1.5.4 Bonferroni-Holm Adjustment

From a theoretical point of view, let us suppose that there are k null hypotheses to be tested and the overall type I error rate is α. This could be the case, for instance, of a comparison between EEG spectrum data for each one of 128 scalp electrodes between two populations by using t-Statistic, either the comparison of two populations voxel by voxel after estimating the source activity by means of sLORETA or other inverse methods available in literature. In this last case the number of comparisons to be made has the size of the voxels employed. The algorithm by Bonferroni-Holm starts by ordering the p-values and comparing the smallest one to α/k. If that p-value is less than α/k, then we may reject that hypothesis and start all over with the same α and test the remaining $(k - 1)$ hypothesis, i.e. order the $(k - 1)$ remaining p-values and compare the smallest one to $\alpha/(k - 1)$. We have to continue doing this until the hypothesis with the smallest p-value cannot be rejected. At that point, we may stop and accept all hypotheses that have not been rejected at previous steps. Hence, the first test will be performed at the actual Bonferroni adjustment, and then the threshold will be gradually decreased on the base of the remaining tests.

2.2 The Experimental Design

In the investigated ecologic situation, as Figure 8 shows, our experimental subjects are asked to comfortably have a seat in front of a computer screen by means of which we present a documentary and a train of three commercial breaks is inserted within, which are our stimuli of interest. The signals gathered during the observation of the documentary will be used to avoid the personal baseline activity. Performing the experiment, subjects are not aware that an interview would be held within a couple of hours from the end of the movie. They are simply told to pay attention to what they watch. In the interview, subjects were asked to recall commercial clips they remember, tell if they are consumers or not of the products advertised and express the degree of pleasantness perceived. In such a way we divide the signals of the population recorded into different datasets.

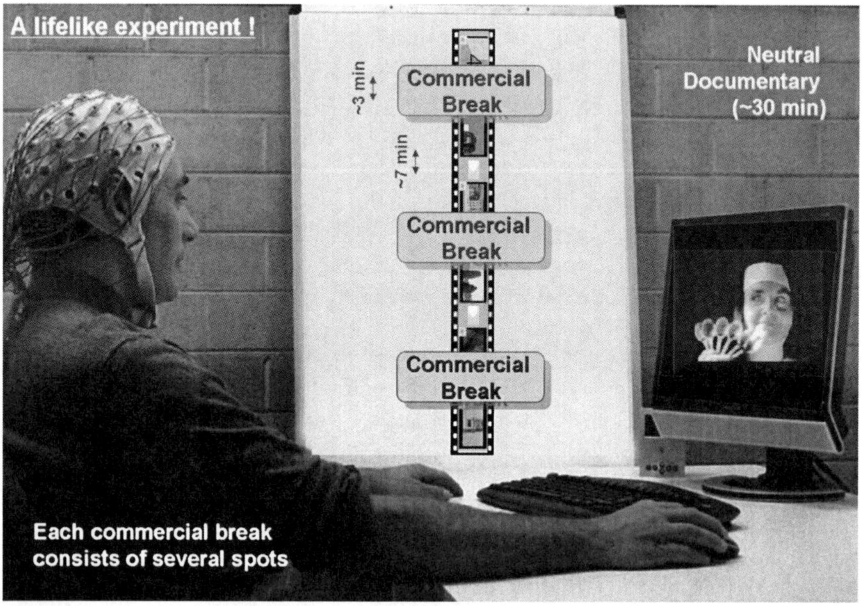

Fig. 4 Picture shows the experimental setup of our neuromarketing studies. The subject is seated in front of a computer screen on which a neutral documentary is running. We inserted a series of commercial breaks within, each consisting of several TV advertisements, which will be our stimuli.

Subjects were comfortably seated on a reclining chair, in an electrically shielded, dimly lit room. A 64-channel EEG system (BrainAmp, Brainproducts GmbH, Germany) was used to record electrical potentials by means of an electrode cap, accordingly to an extension of the 10-20 international system. In the present study, the cortical activity was estimated from scalp EEG recordings by using realistic head models whose cortical surface consisted of about 5000 triangles uniformly disposed. The current density estimation of each one of the triangle, which represents the electrical dipole of the underlying neuronal population, was computed by solving the linear-inverse problem according to the techniques described in literature (Babiloni et al., 2005).

Thus, a time-varying waveform relative to the estimated current density activity at each single triangle of the modeled cortical surface was obtained. Such waveform was then subjected to the time-varying spectral analysis by computing the spectral power in the different frequency bands usually employed in EEG analysis, i.e. theta (4-7 Hz), alpha (8-12 Hz), beta (13-24 Hz) and gamma (24-45 Hz).

In each subject recorded, the statistical significance of the spectral values during the observation of the TV commercials was then measured against the activity evaluated during the observation of the documentary for the same subject. This was obtained by computing a time-varying z-score variable for each subject and for each dipole placed on the cortical mantle in the analyzed frequency band. The mean and the standard deviation for such z-score variable was estimated in the documentary period, while the time-varying values of the spectral power in the theta band during the observation of the TV commercial for each dipole were employed.

In order to present the results relative to the experimental conditions for the entire population, we needed a common cortical representation to map the different activated areas of each subject. For this purpose we used the average brain model available from the McGill University website. In this way we were able to display the cortical areas that are statistically significant activated during different experimental conditions in all subjects analyzed. However, only the statistical significant variation of such spectral power when compared to the documentary period was highlighted in color. The use of the z-score will allow us to have a variable that can be averaged and can be used to synthesize the results of the entire population investigated.

Specifically, cortical activity from EEG scalp recordings was estimated by employing the high resolution EEG technologies (Ding et al., 2005; He et al., 1999; Nunez, 1995; Babiloni et al., 2001; De Vico Fallani et al., 2007) with the use of a realistic head model known as average head model from McGill University. The scalp, skull and dura mater compartments were built by using 1200 triangles for each structures, and the Boundary Element Model was then employed to solve the forward electromagnetic model (Grave de Peralta Menendez, and Gonzalez Andino, 1999). For each subject it was generated the electrodes disposition on the scalp surface, through a non-linear minimization procedure (Astolfi et al., 2008b). The cortical model consists of about 5,000 dipoles uniformly disposed on the cortical surface and the estimation of the current density strength for each dipole was obtained by solving the electromagnetic linear inverse problem according to techniques described in literature (Astolfi et. al 2007a, 2007b; Babiloni et al., 2005) and illustrated in the above sections.

2.3 The Example of How a Mannequin's Head Could "Think"

In order to provide a direct example of the high resolution EEG techniques explained above and to show the importance of applying a statistical correction when dealing with such a tool, in the following we describe a typical neuromarketing experiment in which our experimental "subject" was a mannequin (Vecchiato et al., 2010a). The reason of this recording is to show how, without using a proper statistical analysis, is possible to "retrieve" significant activity also from a foam head . In particular, the mannequin was posed in front of a screen where a video with different contents were displayed, one related to marketing advertisements and the other one related to a documentary. The electrical signals recorded from the high resolution electrode cap employed on the mannequin head was treated in order to extract the spectral power in the typical frequency bands commonly used in the EEG literature. Statistical contrasts between power spectra of the signals collected during the display of the two videos were performed by using different statistical techniques. Figure 9 presents the experimental setup used for the electrical recordings.

2.3.1 Stimuli

The mannequin was posed in front of a screen while projecting a video about a naturalistic documentary. The whole videoclip is 10 minutes long and intermingled with a series of five TV spots related to commercial advertisings, 30 seconds long each. The data we gathered have been segmented in order to separate the signal related to the presentation of the documentary from the one associated to the presentation of commercials. We considered the first dataset as ideally belonging to a resting state, while the second one relates to our experimental condition. We will refer to these signals as REST and TASK datasets respectively. In particular, the TASK dataset is formed by five segments of 30 seconds each while the REST one is formed by a single segment of 150 seconds. These traces have been again segmented in order to obtain 50 trials of 3 seconds each for both REST and TASK conditions. Figure 5 presents the experimental setup employed for the electrical recordings. In particular the mannequin is presented in both experimental conditions: panel A while the documentary is running (REST), panel B during the presentation of a commercial advertisement (TASK).

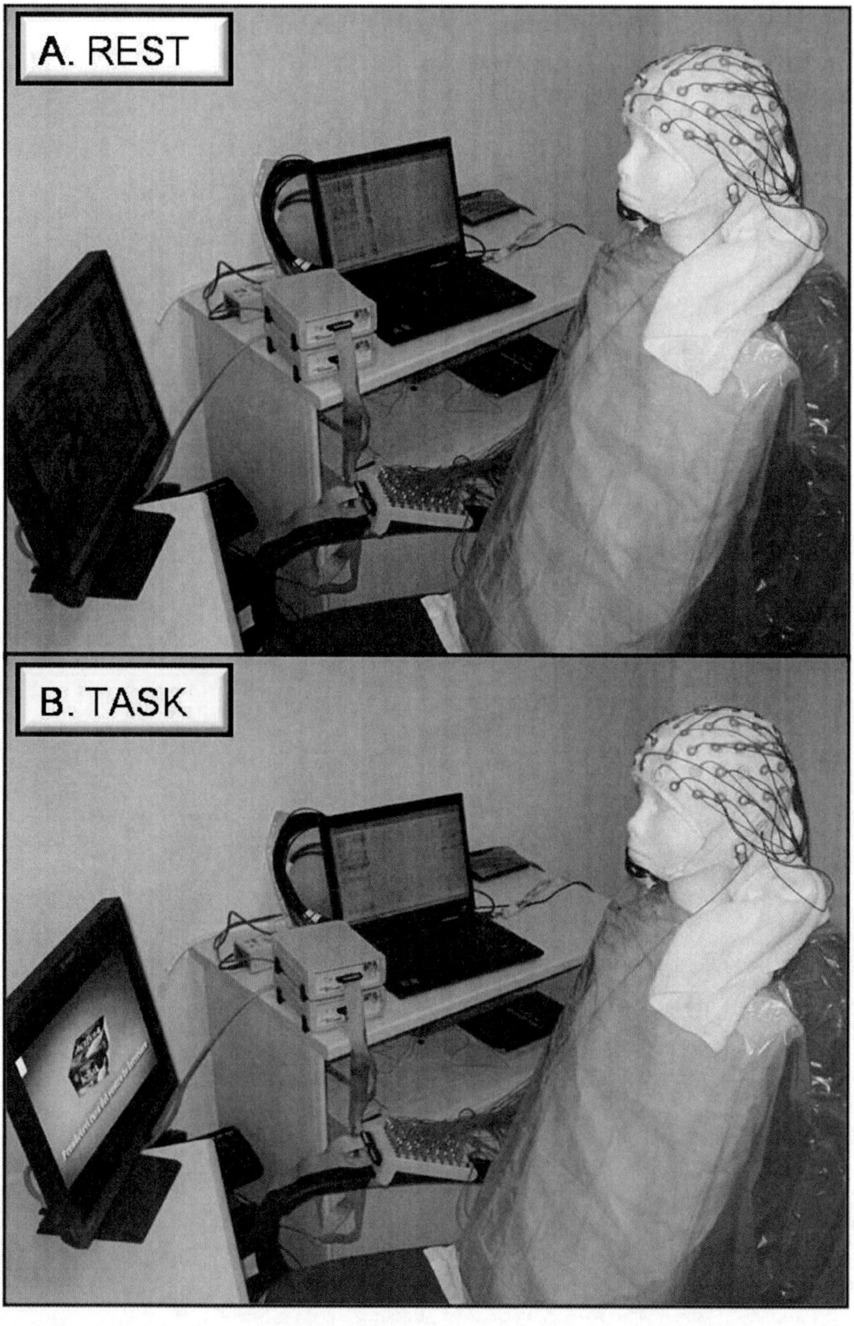

Fig. 5 Figure presents the experimental setup employed for the electrical recordings. In this picture we may see the mannequin in both experimental conditions. Panel A shows the mannequin in front of the screen while the documentary is playing (REST); panel B shows the mannequin while a commercial advertisement is playing on the screen.

2.3.2 Statistical Adjustments

As described in the previous sections, in order to highlight problems which could rise if we employ no statistical protection when performing a high number of statistical multiple comparisons, in the following we present the results obtained by comparing, by means of multiple t tests, the PSD values gathered from the head of a mannequin in two experimental conditions. Our Figures show statistical maps at cortical level, resulting from comparisons with no statistical adjustments, by adopting the False Discovery Rate and the Bonferroni adjustments. The alpha value employed for each t test in the first statistical comparison is 0.05 which will change according to the different adjustment adopted and the level of analysis, namely scalp and cortex. In particular, the alpha values derived from the False

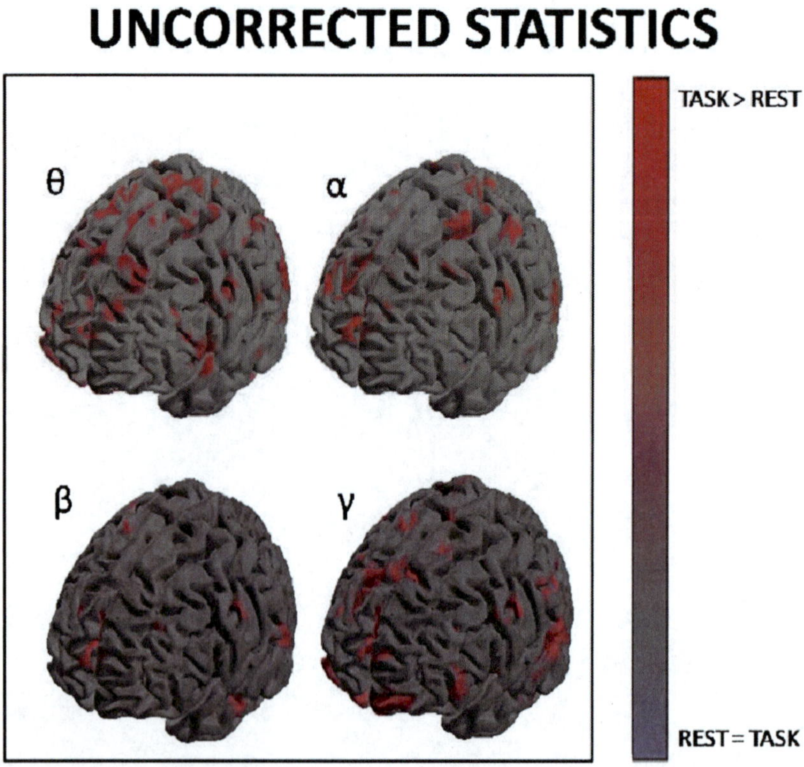

Fig. 6 Figure represents four t-test maps related to the four frequency bands employed for both comparisons at cortical level. Each map refers to the following frequency bands of interest: $\theta = [4, 7]$ Hz, $\alpha = [8, 12]$ Hz, $\beta = [13, 24]$ Hz, $\gamma = [25, 40]$ Hz. Red color highlights scalp and cortical sites in which increased statistically significant activity occurs in the TASK condition when compared to the REST condition in red ($p < 0.05$, no adjustment). Grey colour is used to map scalp and cortical areas where there are no significant differences in the spectral activity between the TASK and REST conditions.

Discovery Rate and the Bonferroni adjustments are $\alpha_{FDR} = 0.0250$ and $\alpha_{Bonferroni} = 3.0516*10\text{-}6$ for the cortical statistics.

For each of the following Figures, the grey color represents the absence of statistical significant electrical activity during the two conditions provided in this experiment, i.e. visualization on the screen of the commercial advertisings (TASK) and the visualization on the screen of the documentary (REST). Instead, the red color highlights the cortical areas in which the electrical activity statistically differs between the two conditions above the 5% chance level, as obtained by the t-Statistic employed. Each picture shows four statistical spectral maps related to the four frequency bands of interest in which the realistic head model is seen from a frontal side perspective.

Fig. 6 presents the statistical significant activity in the two conditions over the used cortical model, at the 5% of threshold employed (uncorrected for multiple comparisons). It is possible to note that such differences are relatively spread over particular areas of the cortex, localized in the left hemisphere, due to the volume conduction effects and the inverse estimation process performed. The picture

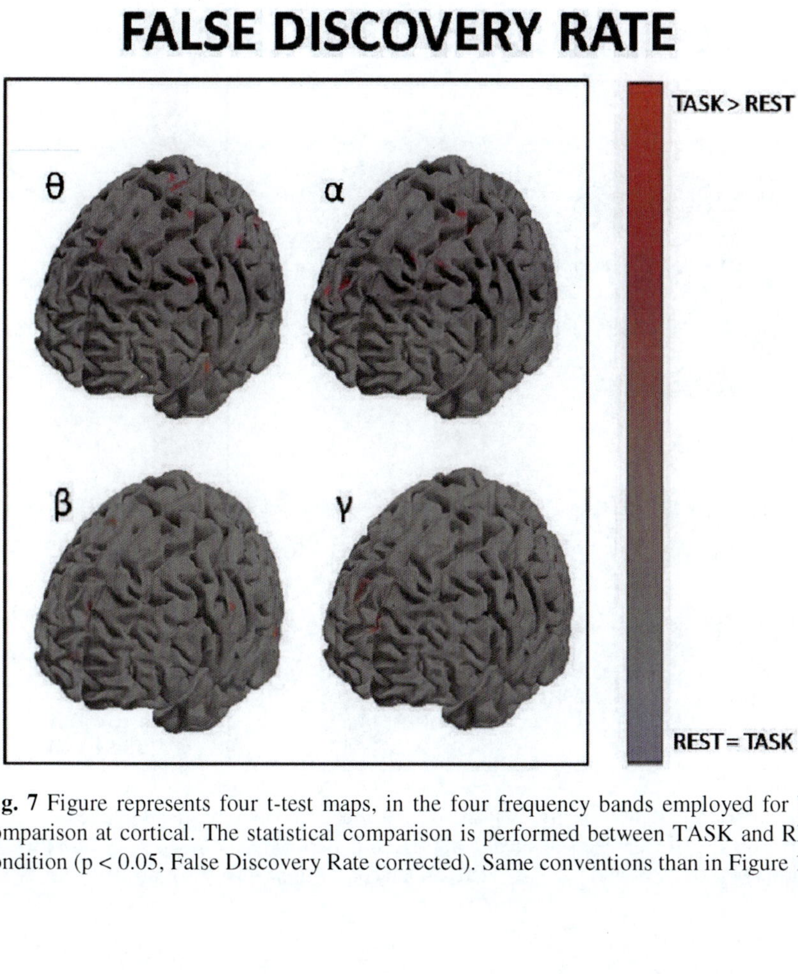

Fig. 7 Figure represents four t-test maps, in the four frequency bands employed for both comparison at cortical. The statistical comparison is performed between TASK and REST condition ($p < 0.05$, False Discovery Rate corrected). Same conventions than in Figure 10.

reports the four frequency bands relative to the uncorrected statistic for the multiple comparisons performed in these experimental conditions.

As briefly reported in the previous section, on the same electrical data treated with the uncorrected statistics, we also applied two other statistical methods, namely False Discovery Rate and Bonferroni adjustment.

In Figure 7 it is possible to appreciate the reduction of the occurrence of the significant differences between TASK and REST conditions after the application of the False Discovery Rate adjustment. In fact, with respect to the same comparison without any statistical adjustment (Figure 6), there are much less spot of activations which fall below the corrected statistical threshold β^*. Again, the red color highlights the existence of a significant statistical difference in the power spectra between the signal estimated over the cortical model in the two analyzed conditions by using the False Discovery Rate. By examining the Figure 7, it can be recognized that, for each frequency band, the spread of the statistical differences is rather reduced when compared to the distribution estimated by employing the uncorrected t-Statistics (Figure 6).

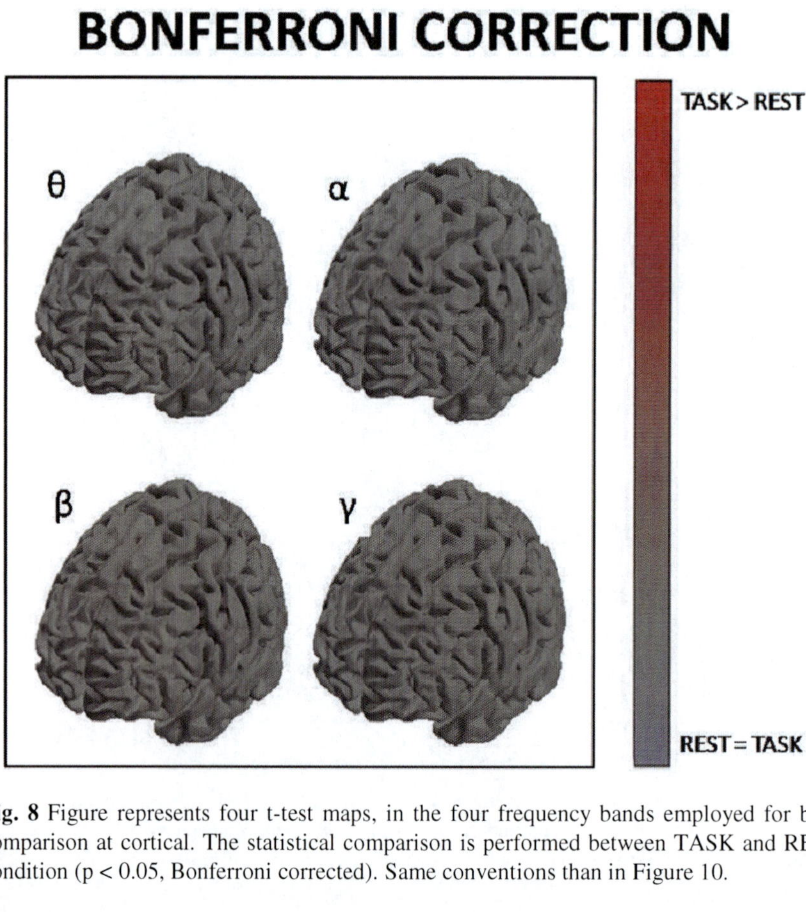

Fig. 8 Figure represents four t-test maps, in the four frequency bands employed for both comparison at cortical. The statistical comparison is performed between TASK and REST condition ($p < 0.05$, Bonferroni corrected). Same conventions than in Figure 10.

The use of Bonferroni adjustment or the Bonferroni-Holm adjustment returned no statistical differences (at 5% of significance) at cortical level (Figure 8). For an easy visualization, we simply present the results obtained by applying the Bonferroni adjustment, being the results of the Bonferroni Holm identical.

Not surprisingly, the use of uncorrected statistic returned an activation of a plausible cluster of voxel within the brain of the mannequin. Such activations, of course, were removed by the use of appropriate statistical techniques. However, it is important to practically demonstrate that the errors could occur by neglecting the adjustment, as in the case here presented. We feel that the illustration of our mannequin, that presents statistical significant differences between the visualization of a commercial advertising when compared to a documentary, have a didactic value. The use of the Bonferroni or Bonferroni-Holm adjustments returned correctly no differences between the signals gathered from the mannequin in the two experimental conditions. However, the uncorrected and the False Discovery Rate where prone to display several differences in spectral power between the signal recorded in the two experimental conditions. A point to be considered, however, is the fact that a disadvantage of using a very conservative statistical adjustment factor such as the Bonferroni adjustment is the increased incidence of Type II error. The experimental paradigm presented here does not permit the exploration of the occurrence of Type II error. As a result, while the Bonferroni adjustment produced favourable results in terms of Type I errors in the present situation, the abundance of Type II error may limit the usefulness of this technique in the occasion in which such error could be relevant for the clinical conclusions to be drawn from the experiment. This has to be kept in mind when such statistical comparisons have to be performed.

Chapter 3
The Track of the Electric and the Magnetic Brain Activity

The most popular brain imaging method adopted in the neuromarketing field is the functional Magnetic Resonance Image (fMRI), a techniques that returns a sequence of images of the cerebral activity by means of the measure of the cerebral blood flow. Although such as images are "static", i.e. they are related to around ten seconds activity, they have a high spatial resolution that no other neuroimaging method can offer. Nowadays, fMRI scanners are currently used in the neuromarketing field and in literature there exist some scientific studies showing the activation of particular cerebral areas during the tasting of a couple of popular drinks such as Coca-Cola and Pepsi (McClure et al., 2004).

It is very well known that the hemodynamic measurements of the brain activity allow a level of localization of the activated brain structures on the order of few cubic mm, being capable to detect activations also in deep brain structures such as amigdala, nucleus accumbens etc. However, the lack of time resolution, due to the delay of the cerebral blood flow's increment after the exposition to the stimuli, make the fMRI unsuitable to follow the brain dynamics on the base of its subseconds activity.

However, there are other brain imaging techniques that allow to follow on a millisecond base the brain activity during the exposition to relevant marketing stimuli. Such techniques are the electroencephalography (EEG) and the magnetoencephalography (MEG). One problem of such brain imaging techniques is that the recorded electrical or magnetic cerebral signals are mainly due to the activity generated on the cortical structures of the brain. In fact, the electromagnetic activity elicited by deep structures (usually advocated for the generation of emotional processing in humans) is almost impossible to gather from usual superficial EEG electrodes or MEG sensors (Nunez, 1995; Urbano et al., 1998). To overcome this problem, high resolution EEG technology has been developed to enhance the poor spatial information content of the EEG activity in order to detect the brain activity with a spatial resolution of a squared centimeter and the unsurpassed time resolution of milliseconds [Nunez, 1995; Bai et al., 2007; He et al., 1999; Dale et al., 2000; Babiloni et al., 2005].

On the other hand, by using the MEG signals it is also possible to estimate the activity at the level of brain voxels by using the Magnetic Field Tomography

G. Vecchiato et al.: *Neuroelectrical Brain Imaging Tools*, BIOSYSROB 3, pp. 33 – 44.
DOI: 10.1007/978-3-642-38064-8_3

(MFT) that it is the name given to the resulting method which perform the estimation of brain activity (Ribary et al., 1991). MFT solutions can scrutinize brain function at multiple spatiotemporal scales. In the spatial domain, the range covers details of a few millimeters allowing to map almost the entire brain. In the time domain, events can be analyzed at timescales from a fraction of a millisecond to minutes and hours (Ioannides, 2006).

It is worth of note that EEG and MEG techniques, while exhibit a remarkable time resolution and a passable spatial resolution, have drastically different costs for the marketing research. In fact, MEG technology uses liquid helium and needs special shielded structures in order to record the tiny brain magnetic signals produced of the order of fTesla. On the contrary, EEG devices are relatively inexpensive, robust and even wearable by the subject, making such EEG technology of interest for the evaluation of marketing stimuli (for a review on EEG and MEG in neuromarketing see Vecchiato et al., 2011a).

3.1 Electroencephalographic Temporal Patterns of Cortical Activity

The purpose of this section is to illustrate the potential of the high resolution EEG techniques when applied to the analysis of brain activity related to the observation of TV commercials. In particular, we would like to describe how by using appropriate statistical analysis it could be possible to recover significant information about cortical areas engaged by particular scenes inserted within the TV commercial analyzed.

In order to do that, we recorded a series of healthy subjects with high resolution EEG techniques during the observation of a documentary in which an interruption was generated. The subjects were not aware of the aim of the study. The brain activity was evaluated in both time and frequency domains by solving the associate inverse problem of EEG with the use of realistic head models (Astolfi et al., 2009).

Cortical activity estimated during the observation of the TV commercial was then compared with the brain activity computed in the analyzed population during the observation of the documentary.

3.1.1 The Experimental Scenario

The dataset used to obtain the results presented in the present Chapter is composed by EEG registrations of thirteen healthy subjects (mean age 30±4 years) watching a documentary of 30 minutes intermingled by a TV commercial (see Astolfi et al., 2008c). Each subject is exposed to the observation of a same documentary. The subjects were not aware of the aim of the recording, and they only knew to pay attention to the material showed on the screen during the entire 30 minutes. The TV commercial, whose length was 30 seconds, was inserted at the middle of the documentary. Such commercial was realized for a popular brand of beer in Italy, that was on-air on the national TV channels on the days in which the experiment was realized. After the EEG registration each subject was recalled in laboratory where an interview was performed. In such interview, the subjects were asked if they usually drink beer or light alcohol at least once per week. If yes, subjects were considered within the dataset of “drinkers” in opposition to the dataset of “no drinkers”. In order to increase the sensitivity of the analysis performed, only the EEG spectral analysis for the “drinkers” was analyzed and presented here.

The hypothesis was that the TV commercial could be better followed by a class of subjects who usually drink beer instead that from other “non-drinkers” subjects.

3.1.2 Statistical Maps of Cortical Activity

In this work, thirteen subjects were collected each exposed to the observation of a documentary in which we inserted a TV commercial of a popular Italian beer. After the EEG recording we interviewed all the experimental subjects in order to obtain information about their drinking behavior. Of the thirteen subjects recorded in this experiment, only seven are “drinkers”. Hence, the successive analysis and results are presented for seven of such subjects. We summarized all results for the “drinkers” group in a series of Figures showing the statistically significant differences of cortical activation concerning this dataset in the theta frequency band (4-7 Hz). Data regarding the alpha frequency band (8-12 Hz) were equivalent to the theta band and for this reason not shown here. Our Figures are formed by a series of subsequent panels each containing two images: the upper one represents a frame of the TV commercial while the lower one displays the corresponding mean brain activity. In particular, the image at the bottom of the panel shows four different views of the average brain model organized in two rows: the upper row comprises the front and left perspective while the lower one the rear and right brain perspective. The temporal axes beat the time of the commercial. In Figure 9 we present a first series of 7 film segments spanning the whole length of a certain TV spot. Frames are taken each 5 seconds from the beginning of the clip.

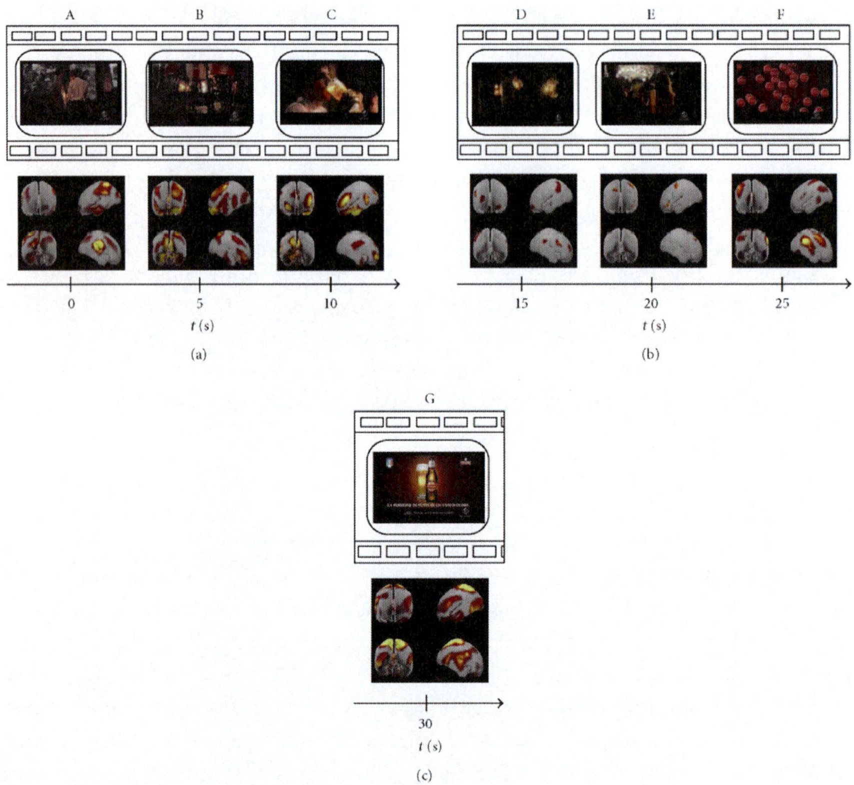

Fig. 9 Tracking of the mean cortical activity of the group of "drinkers" in the theta frequency band spot. The statistical significant activity in this population is shown in seven panels (A-G), each representing subsequent film segments of a TV spot with corresponding brain activity. Temporal axes beats the spot time every 5 seconds: in this way panel A represents the first frame of the commercial while panel G shows the last one. This example illustrates how is possible to track human cortical activity by means of the High-Resolution EEG technique.

In such a way panel A represents the first frame of the commercial while panel G shows the last one. By examining this strip it results evident how the temporal evolution of the mean cortical activity changes according to the images viewed by the subjects. In particular, an enhancement of cerebral activity is suggested by the result of the application of the statistic tests at the beginning and at the end of the videoclip presented. In fact, from the lower row of the Figures, it is possible to observe how in the middle film segments very restricted areas provide statistically significant differences when compared to the ones watched at the beginning and at the end of the commercial. This drastic change of activity is more evident in Figure 10.

The present Figure is composed by 3 panels representing the first (panel A), the middle (B) and the last (C) frame of the TV spot respectively. The corresponding mean cortical activity completes each panel of the Figure. By observing these

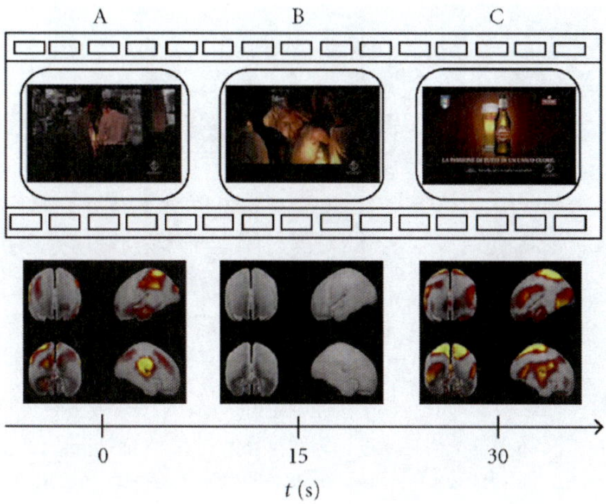

Fig. 10 Tracking of the mean cortical activity of the group of "drinkers" in the theta frequency band spot. The statistical significant activity in this population is shown in 3 panels (A-C), each representing subsequent film segments of a TV spot with corresponding brain activity. Temporal axes beats the spot in correspondence of the beginning (A), the middle (B) and the end (C) of the entire film sequence.

three images is clear how the middle part of the commercial is characterized by cerebral zones displaying no statistical differences across ROIs, while there are two peaks of activity at the beginning and at the end of the clip.

The analysis of the temporal evolution of the brain activity has been performed even on shorter intervals in order to track its variations in closer time instants.

Subsequent Figures 11, 12 and 13 follow the cerebral activity with a higher temporal resolution.

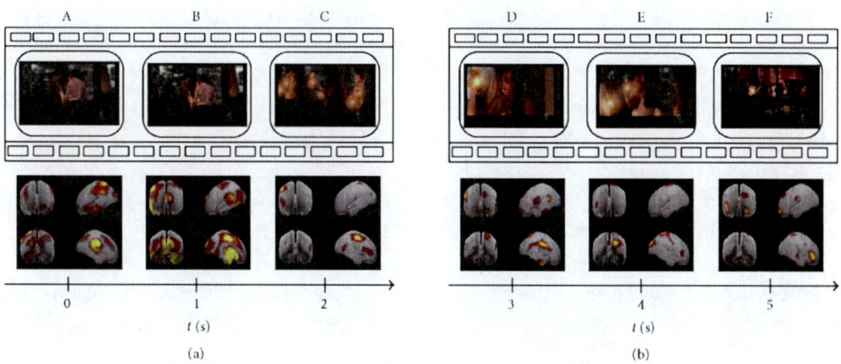

Fig. 11 Tracking of the mean cortical activity in the theta frequency band of the first 5 seconds of the commercial spot. The statistical significant activity in this population is shown in six panels (A-F), each representing subsequent film segments of a TV spot with corresponding brain activity. Temporal axes beats the spot time every second: in this way panel A represents the first frame of the commercial while panel F shows the film segment shown after 5 seconds from the beginning.

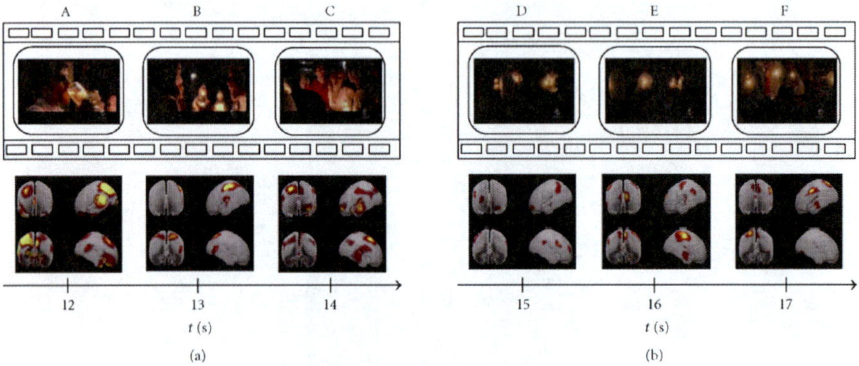

Fig. 12 Tracking of the mean cortical activity of seven drinkers in the theta frequency band of the central 5 seconds of the commercial spot. The statistical significant activity in this population is shown in six panels (A-F), each representing subsequent film segments of a TV spot with corresponding brain activity. Temporal axes beats the spot time every second: in this way panel A represents the film segment after 12 seconds from the beginning of the commercial; panel F shows the film segment after 17 seconds.

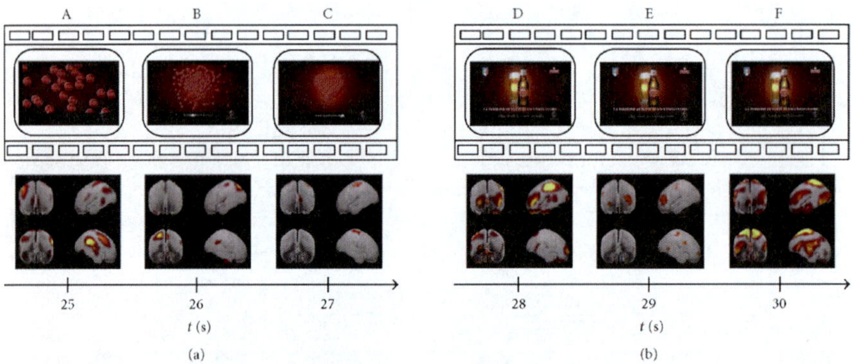

Fig. 13 Tracking of the mean cortical activity of the last 5 seconds of the commercial spot. The statistical significant activity in this population is shown in six panels (A-F), each representing subsequent film segments of a TV spot with corresponding brain activity. Temporal axes beat the spot time every second: in this way panel F represents the last film segment of the commercial; panel A shows the film segment after 25 seconds from the beginning.

These examples show how it is possible to catch statistically significant differences in the activation of cortical areas even reducing the time interval of interest.

3.1.3 Differences between Consumers and Non-consumers

A different pool of data is formed by the activity associated to subjects who affirmed to use the item advertised. This group has been named CONSUMERS.

Conversely, we also distinguished the group comprising all the cerebral activity of subjects who answered telling to not use the product advertised. We referred to this dataset as NON CONSUMERS (Vecchiato et al., 2012a).

In Figure 14 we present two series of film segments spanning the length of a particular TV commercial in which we illustrate the statistically significant differences of cortical activations concerning both the CONSUMERS (panel A) and NON CONSUMERS (panel B) groups in the theta band.

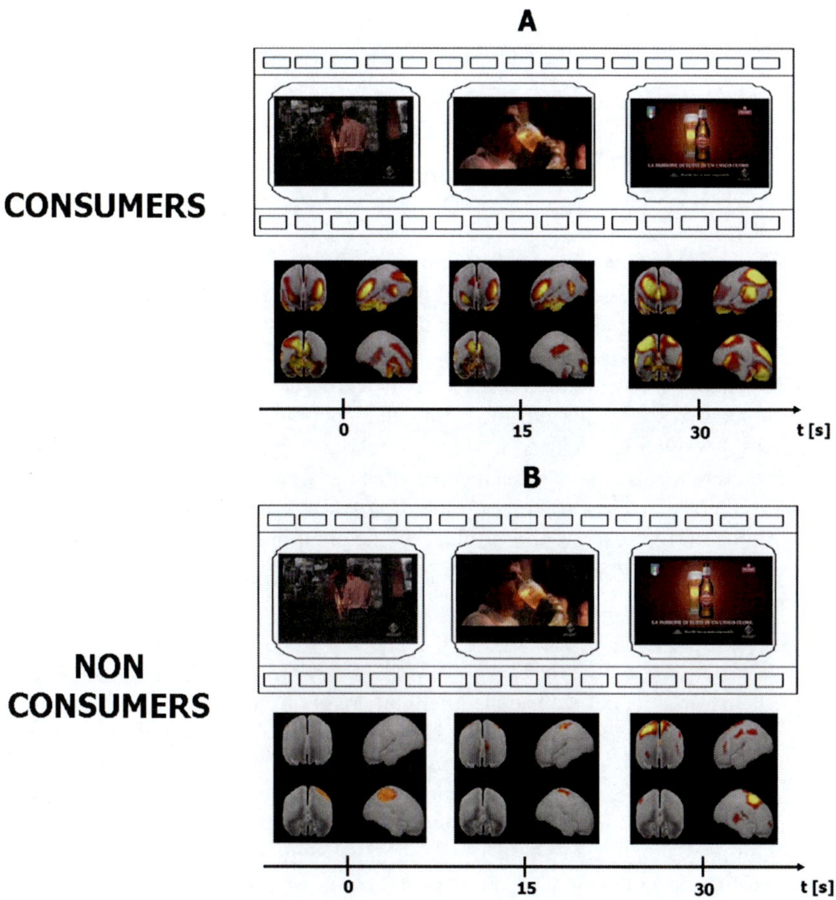

Fig. 14 Brain activity related to the observation of a commercial advertisements for the two experimental groups CONSUMERS and NON CONSUMERS, panel A and panel B of the picture respectively. Each panel two strips: the upper one represents three frame segments of the commercial of interest at time instant 0, 15 and 30 seconds; similarly, the lower strip shows the related brain activity for each time instant. The average cortex model is seen from four different perspectives (frontal, right, posterior and left side from the left to right and from the top to the bottom respectively). The red and yellow colors highlight the statistically significant increases of theta activity during the observation of the commercial when compared to the observation of the documentary.

In particular, the panel A of the presented picture shows how the cerebral activity of the CONSUMERS is characterized by wide cortical activations in several key frames, located in cortical areas mostly involving frontal and parietal lobes. Conversely, in the same key frames, the cortical activations related to the NON CONSUMERS (panel B) present less activations when contrasted with the observation of the documentary. By comparing the two experimental groups, it is possible to observe that this advertisement elicited an increase of PSD in the theta band for the only group of CONSUMERS. Since such activity is linked to the memorization process, it could be suggested that the ad presented is more adequate to promote the item in the CONSUMERS population than in the other one.

3.2 Magnetoencephalographic Temporal Patterns of Cortical Activity

In the context of the neuroelectrical imaging, the research team of Sven Braeutigam (2005, 2004, 2001) has employed the MEG in order to study the temporal relationship of cerebral areas involved in consumers' choices when they have to make decisions among different items within a laboratory context.

In their studies they wanted to analyse the cerebral behaviour by distinguishing male subjects from females during a simulated shopping.

Cerebral activations induced by multiple choices reflect the level of familiarity or the preference that a particular experimental subject had with the presented products. These factors can be considered by taking into account the relationship between the current choice of a product on the shelf and the relative frequency of choice and usage of that product in the past.

3.2.1 Product Choice

As far as the product choice is concerned, the main observation came to light from these studies presents the consumer's choice like a complex sequence of cerebral activations that greatly differ according to the consumer's sex and to the probability of choice. From a behavioural point of view, choices with a high probability were faster than those less predictable. This can be interpreted by supposing that in the case of more difficult choices the cortical activities are more complex than those simple to make. As illustrated in Figure 15, they distinguished two distinct cerebral paths. The first one is referred to predictable choices, i.e. associated to products that the experimental subject already used in the past or said to prefer; the second one is related to unpredictable choices, i.e. associated to unfamiliar products to the subject. In the experiment performed by Brautigam, the first stage in the decisional process has been individuated around 100 ms after the stimulus onset with an activity located in the occipital cortex. At that stage of decision (Working memory, W) the subject compared the product to choose with the list of products seen before by involving the working memory, although this early component is also usually related to the processing of sensory percepts.

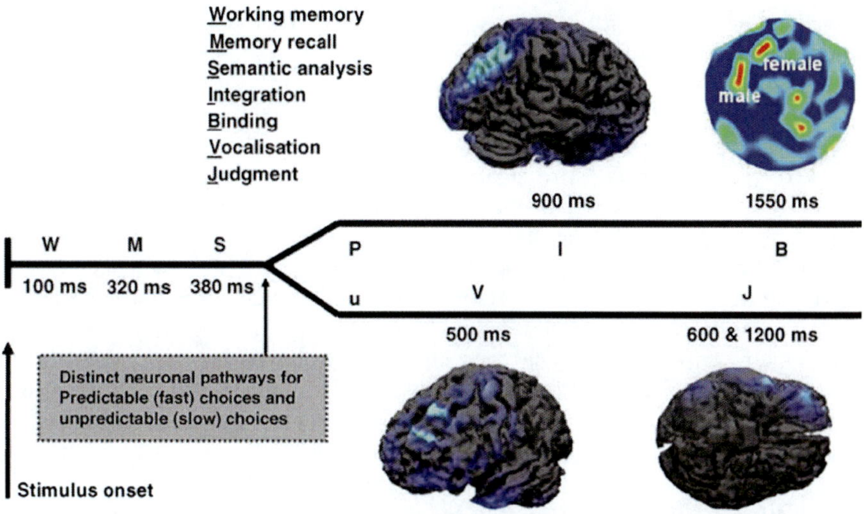

Fig. 15 Cortical activations associated to the decisions of an experimental subject. Predictable choices (P) are the ones related to familiar items which have been often bought or used in the past, compared with the unpredictable ones (u). Label letters indicate the different stages of the trial, namely Working memory, Memory recall, Semantic analysis, Integration, Binding, Vocalization, Judgement, as the legend indicates as well. Cortical maps present the brain areas activated (blu, lower activity; red higher activity) during the different decision stages in the frequency range from 30 to 40 Hz. Activations over the left prefrontal cortices are gender-related, as the bi-dimensional map shows (high perspective, nose up). Reproduced with permission from (Braeutigam, 2005).

The sequence of cortical activations observed in the experimental subjects continues with two neuronal stages partially correlated (Memory recall M; Semantic analysis, S) which can be observed between 280 and 400 ms after the beginning of the decisional process. In this period the selective attention of the subject is oriented towards images of products to identify, classify and compare with those stored in the memory related to the preferred products and brands. This memory can involve the past experience to have bought the particular item or to have watched the commercial of the specific brand.

3.2.2 Gender Differences

The cerebral activation in this time interval differs between men and women. In particular, around 400 ms after the stimulus onset, female subjects presented a stronger activation with respect to the males in the left parieto-occipital lobe of the brain while males presented a stronger cortical activity in the right temporal lobe, although authors do not show such as results. These differences connected with the sex of the subjects characterize both the stage of choice of the product and its discrimination. These observations suggest that, at this temporal stage, women tend to employ a strategy based on the knowledge of the product to buy, while men tend to act according to a spatial memory strategy (Kimura, 1996). In addition, we could also say that men and women act according different semantic interpretation of the stimuli observed.

After 500 ms from the beginning of the decisional process, two patterns of cortical activations can be identified according to the predictability of the choices adopted by the subjects. In particular, as to the predictable choices (P), we can observe a strong activation in the right parietal areas around 900 ms after the beginning of the experiment (Integration, I). In later time latencies, predictable choices of products recall strong MEG oscillations in the frequency band between 30-40 Hz in the left prefrontal cortex (Binding, B). Parietal cortex receive inputs from many cortical areas since it is involved in the spatial integration of sensorial information. Differences in the cortical activity between men and women can strengthen the hypothesis of two different groups of strategies. On the contrary, unpredictable choices (u) generate a strong activation in the right inferior frontal cortex (Vocalisation, V), at a latency of around 500 ms, and in the left orbitofrontal cortex (Judgement, J) between 600 and 1200 ms after the stimulus presentation. In the case V, the cortical patterns are consistent with the activity in the Broca's area, which is involved in the spoken language, also active during the observation of videoclips. Hence, the cortical activity at this latency may indicate a tendency to vocalize brands, as a part of strategy which helps in the decision when it is difficult. The activity in the orbitofrontal cortex (J) can be explained by stating that during an unpredictable choice we have to evaluate the outcome in terms of convenience. Overall, these results present a complex neuronal network which is active during a simple decisional process connected to the purchase of a product. The generation of a choice is considered as an information processing which can be highly influenced, sensible to the complexity of the decision to make and to the rush in which the decision is made along with many other factors.

3.2.3 TV Advertising

A strong involvement of parietal areas during the observation of the TV commercials with an affective and cognitive content was also noted in a previous study, performed by using sophisticated MEG recordings (Ioannides et al., 2000). In that study, cognitive frames elicited a stronger activity in the parietal areas and superior prefrontal cortex while the observation of the affective ones is correlated with the activation of the orbitofrontal and retrosplenial cortex, amygdala and brainstem. The magneto field tomography (MFT) results showed an increasing activity during the observation of cognitive stimuli rather than affective commercials in parietal and superior prefrontal areas. These regions are known to be associated with executive control of working memory and maintenance of highly processed representation of complex stimuli (Summerfield and Mangels, 2005). Although the affect related activations are more variable across subjects, these findings are consistent with previous PET and fMRI studies (Cahill et al., 1996; Maddock, 1999; Grabenhorst et al., 2008; Babiloni et al., 2003) showing that stimuli with affective content modulates activity in the orbitofrontal and retrosplenial cortex, amygdala and brainstem.

3.3 The Capabilities of the High Resolution EEG in Neuromarketing

Thanks to the high resolution EEG techniques we tracked subjects' brain activity during visualization of the commercials: in such manner it is possible to obtain a global measure of the reconstructed cortical signals by means of a simple graphic tool which allows us to distinguish the activity of different cortical areas. The above mentioned results allow us to comment temporal and spatial events observed.

In fact, the observed phenomena suggest an active role of the prefrontal and parietal areas in the coding of the information possibly retained by the users from the TV commercials. A statistical increase of EEG spectral power in the prefrontal (namely Brodmann areas BA 8, 9) and parietal areas is in agreement with the suggested role of these regions during the transfer of sensory percepts from short-term memory to long-term memory storage. The results suggest a strong prevalence of a 'common' prefrontal bilateral (involving BA 8 and 9) activity in all the subjects analyzed during the observation of the TV commercials. In addition a stronger involvement of the left frontal areas has been noted, in agreement with the HERA model (Tulving, 1994) in which such hemisphere plays a decisive role during the encoding phase of information from the short-term memory to the long-term memory, whereas the right hemisphere plays a role in the retrieval of such information. It must be noted, however, that the role of the right cortices in storing images has been also recognized for many years in neuroscience (Braeutigam et al., 2002, 2005).

As presented in previous works performed both with EEG analysis and MEG recordings (Astolfi et al., 2008c; Ioannides, 2000), the observed phenomena suggest an active role of the prefrontal and parietal areas in coding of the information that will be retained by users from the TV commercials. In particular, activations of these cortical areas can be associated with attentional and memorization processes. As shown in the previous Figures, peaks of activity emerge at the beginning and at the end of clip (Figures 9, 10). In these periods subjects' attention is more focused on what he/she sees, in particular when they watch scenes showing meeting moments (such as panels A, B in Figure 11) and the advertised product (panel D and F of Figure 9). Instead, in the middle of the TV clip, we observed a peak of activity only when subjects watch a person utilizing the advertised product (such as a beer in panel A of Figure 12). These processes could reflect memorization of significant frames' sequence which would help the subject to understand the whole video clip and messages provided. Climax of this elaboration will be achieved in the last film segments of the sequence when the meaning of the commercial will be completely understood (last panel of Figure 13).

The present result intends to stress the useful properties of the High Resolution EEG technologies. In particular this tool is able to help us in observing and analyzing the temporal trend of the cortical activities thanks to a high temporal and spatial resolution. These features allow us to distinguish with a certain precision changes of activation of ROIs corresponding to different cortical areas, by means of a graphical representation on an average brain model. Our analysis focused attention on tracking of human brain activity with different time resolution, all offering the same spatial resolution able to discriminate activation's intensity of Brodmann areas.

The reconstruction of the cortical activity by means of the high resolution EEG technique and by combining the above statistic treatment of our data, allowed us to track subjects' brain activity during visualization of the commercials. In such a way for each film segment of a clip it was possible to distinguish cortical areas that were significantly activated when compared to the observation of the documentary. This could be useful in the evaluation of the cortical responses to particularly type of visual solicitations, performed by film or commercial clips, that at the moment is a field largely unexplored by neuroscience.

Chapter 4
Cortical Correlates of Cognitive and Emotional Processes

In the present Chapter we are going to illustrate the details of an experiment which helped us in understanding the cortical correlates of cognitive and emotional processes during the observation of TV commercials. Particularly, we investigated the cerebral activity of fifteen healthy subjects by gathering and analysing the EEG, GSR and HR activity while they were watching a documentary intermingled with a series of advertisements. After the data acquisition, subjects were asked to answer a questionnaire which allowed us to group the data according their preference of pleasantness (LIKE/DISLIKE groups) and performance of memorization (RMB/FRG groups). In the different sections of the Chapter, methods and results of the work (Vecchiato et al., 2010b) are illustrated and commented.

4.1 Recording and Analysis of the Cerebral Activity

The cerebral activity was recorded by means of a portable 64-channel system (BE+ and Galileo software, EBneuro, Italy). Informed consent was obtained from each subject after explanation of the study, which was approved by the local institutional ethics committee. All subjects were comfortably seated on a reclining chair, in an electrically-shielded, dimly-lit room. Electrodes positions were acquired in a 3D space with a Polhemius device for the successive positioning on the head model employed for the analysis. Recordings were initially extra-cerebrally referred and then converted to an average reference off-line. We collected the EEG activity at a sampling rate = 256 Hz while the impedances kept below 5 kΩ. Each EEG trace was then converted into the Brain Vision format (BrainAmp, Brainproducts GmbH, Germany) in order to perform signal pre-processing such as artefacts detection, filtering and segmentation. Raw EEG traces were first band pass filtered (high pass = 2 Hz; low pass = 47 Hz) and the Independent Component Analysis (ICA) was then applied to detect and remove components due to eye movements, blinks, and muscular artefacts. These EEG traces were then segmented to obtain the cerebral activity during the observation of the TV commercials and that associated to the REST period.

G. Vecchiato et al.: *Neuroelectrical Brain Imaging Tools*, BIOSYSROB 3, pp. 45–62.
DOI: 10.1007/978-3-642-38064-8_4

Since we recorded such activity from fifteen subjects, for each proposed advertisement we collected fifteen trials which have been grouped and averaged to obtain the results illustrated in the following sections. This dataset has been used to evaluate the cortical activity and calculate the power spectral density (PSD) for each segment.

Besides, we separately analysed two more traces derived from the previous one. Each EEG trace has been band pass filtered two more times in order to isolate the only spectral components in the theta band and those located between the beta and gamma band, that we call in the following the extended beta band (high pass = 13 Hz; low pass = 45 Hz; beta) from the whole EEG spectrum. All segments were exported in binary format and then converted for further data processing performed with in-house MATLAB software.

These filtered theta and extended beta filtered traces have been employed to calculate the Global Field Power (GFP; Lehmann and Skrandies, 1980) for each segment, then converted in z-scores in order to extract cerebral indexes following described. Since for the phenomena we would like to investigate a clear role of the frontal areas have been depicted (Summerfield and Mangels, 2005; Werkle-Bergner et al., 2006) we used the frontal electrodes to compute the GFP indexes used in the following of this study. The filtered EEG signals were subjected to the computation of the Global Field Power by taking into account the signals that comes from the following frontal and prefrontal electrodes of the 10-10 International System: F3, F4, AF3, AF4, F7, AF7, F8, AF8, Fz, AFz.

4.1.1 Contrast of Cortical Spectral Maps

We initially calculated statistical spectral maps for each subject, each TV commercial, in the four frequency bands. Since we transformed the PSD data into Z variables, it has been possible to group the single subjects activities according to the answers which they gave during the interview. In this way the cerebral activity recorded during the observation of the advertisements has been considered as belonging to the groups RMB and FRG or LIKE and DISLIKE. The cortical maps depicting the statistical contrasts between RMB and FRG conditions as well as the LIKE and DISLIKE conditions were then generated for each frequency band considered.

All the statistically-activated areas of each subject were mapped on a common cortical representation through such transformation. For display purposes, we represented the results obtained from the average brain model created with the BRAINSTORM software freely downloadable from internet (http://neuroimage.usc.edu/brainstorm/). In particular, the average brain model was used to display the cortical areas that are statistically significantly activated during the different experimental conditions in all the subjects analyzed.

The EEG signals gathered during the observation of the commercial ads were subjected to the estimation of the cortical power spectral density by using the techniques described in the Methods section. In each subject, the cortical power spectral density were evaluated in the different frequency bands adopted in this study and contrasted with the values of the power spectral density of the EEG during the observation of the documentary through the estimation of the z-score.

These cortical distributions of the z-scores obtained during the observation of the commercials were then organized in two different populations: the first one was composed by the cortical z-scores relative to the observation of commercial videos that were remembered during the interview (RMB group), while the second was composed by the cortical distribution of the z-scores relative to the observation of commercial videos that were forgotten (FRG group). A contrast has been made between these cortical z-score distributions of these two population, and the resulting cortical distributions in the four frequency bands highlight the cortical areas in which the estimated power spectra statistically differ between the populations.

Figure 16 presents four cortical maps, in which the brain is viewed from a frontal perspective. The maps are relative to the contrast between the two population in the theta (upper left), alpha (upper right), beta (lower left) and gamma (lower right) frequency bands. The color scale on the cortex coded the statistical significance: where there are cortical areas in which the power spectrum does not differ between the two populations, a grey color was employed. The red color was employed when the cortical areas present a statistically significance power spectral activity greater in the population that remembered the commercial videos (RMB) with respect to the other, while the blue color coded the opposite situation (i.e. the power spectral activity in the group that forget the commercial videos is greater with respect to the brain activity in the group that remembers the ads).

Figure 16 presents an increase of cortical activity in the theta band that it is prominent on the left pre and frontal hemisphere for the RMB group. The statistical significant activity in the alpha frequency band for the RMB group is still increased in the left hemisphere although there are few zones in the fronto-central and right prefrontal hemisphere where the cortical activity was prominent for the FRG group. In the beta band there are spots of significant increase of cortical activity for the RMB group when compared to the FRG group on the left pre and frontal hemispheres, while increase of cortical activity in the FRG group is scarcely present on the right hemisphere.

Finally, in the gamma band is observed a significant increase of cortical activity in a large zone of the pre and frontal hemispheres in the RMB group when compared with the FRG one.

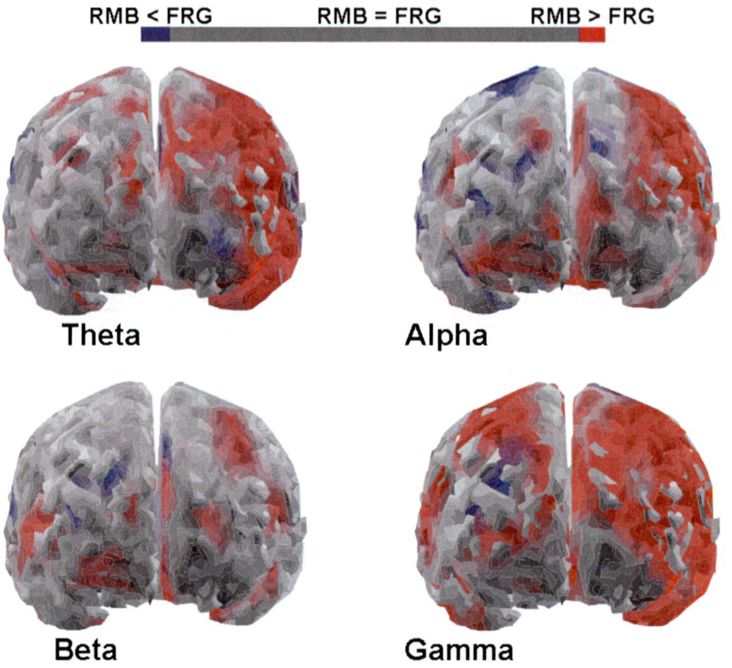

Fig. 16 Figure presents four cortical z-score maps, in the four frequency bands employed. Colour bar represents cortical areas in which increased statistically significant activity occurs in the RMB group when compared to the FRG group in red, while blue is used otherwise ($p < 0{,}05$ Bonferroni corrected). Grey colour is used to map cortical areas where there are no significant differences between the cortical activity in the RMB and FRG groups.

Figure 17 presents the contrast between the LIKE and DISLIKE groups in the four frequency bands considered in this analysis. Same convention of Figure 16 is used. The significant increase of the frontal activity in the theta band is clearly visible (in red) in the LIKE group when compared to the DISLIKE one, in the upper left part of the Figure 17. Scattered increased of cortical activity on the left hemisphere is also present in the DISLIKE group (in blue). In the alpha frequency band (upper right of the Figure 17) significant increase of cortical activity is present on the left hemisphere and on the orbitofrontal right hemisphere in the LIKE group when compared to the DISLIKE one. The cortical activity in the beta band is greater in the DISLIKE group in the prefrontal left areas when compared to the LIKE group (lower left of Figure 17), while the gamma frequency band (lower right of the Figure 17) presents a statistical increase of the activity of the pre and orbito-frontal cortical areas rather bilaterally for the LIKE group when compared to the DISLIKE one.

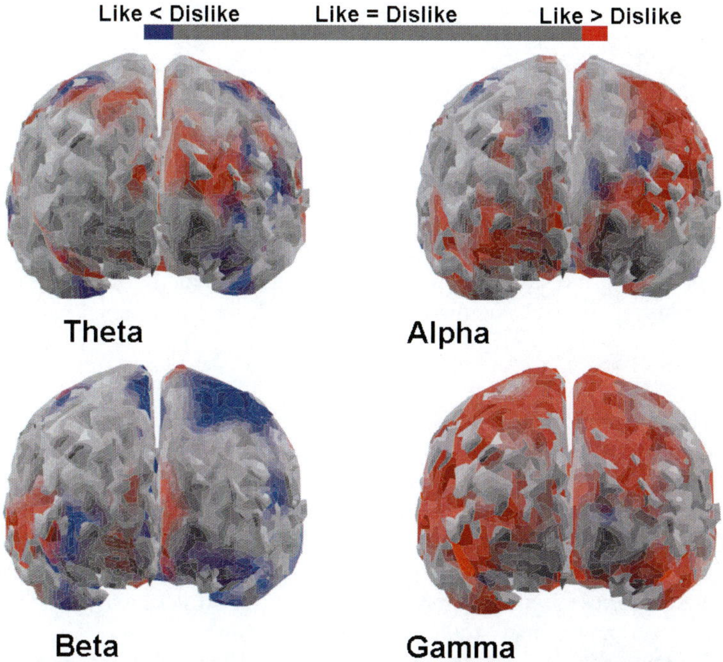

Fig. 17 Four cortical maps for the groups LIKE and DISLIKE. Same conventions than in the previous Figure except the use of red colour to code the cortical areas in which the brain activity of the LIKE group is significantly higher than the activity of the DISLIKE group. The blue colour is used to map brain areas in which the activity of the DISLIKE group is significantly higher than the activity of the LIKE group.

The analysis of the statistical cortical maps in the different conditions (RMB vs FRG and LIKE vs DISLIKE) suggested that the left frontal hemisphere was highly active during the RMB condition, especially in the theta and gamma band, while the activity of the brain is greater in the LIKE condition than in the DISLIKE except that in beta band, being the activity in the LIKE condition for the gamma band rather symmetrical. These results are in agreement with different observations on the RMB condition performed in literature (Summerfield et al., 2005; Werkle-Bergner et al., 2006 Astolfi et al., 2008a). In addition, the results here obtained for the LIKE condition are also congruent with other observations performed with EEG in a group of twenty subjects during the observation of pictures from the International Affective Picture System (IAPS, Aftnas et al., 2004). Such observations indicated an increase of the EEG activity in the theta band for the anterior areas of the left hemisphere. It is worth to note that there were methodological differences between the Aftanas and colleagues study and the present one, mainly related to the use of different material as stimuli, and different processing algorithms. Nevertheless, the convergence of these results, obtained in the "naturalistic" conditions of the observation of commercial videos within the documentary with those of more controlled memory and affective tasks, deserve attention.

Taken together, the results indicated the cortical activity in the theta band on the left frontal areas was increased during the memorization of commercials, and it is also increased during the observation of commercials that were liked by the subjects. These results are in agreement with the role that has been advocated for the left pre and frontal regions during the transfer of sensory percepts from the short term memory toward the long-term memory storage by the HERA model (Tulving et al., 1994). In fact, in such model the left hemisphere plays a key role during the encoding phase of information from the short term memory to the long term memory, whereas the right hemisphere plays a role in the retrieval of such information.

4.1.2 The Global Field Power

The use of the Global Field Power (GFP) has a very long tradition in summarizing the brain activity in particular scalp areas or over the entire scalp surface (Lehmann and Skrandies, 1980). Here, we used the GFP as an indicator of the frontal brain activity in two particular frequency bands; theta and an extended beta band. The choice for these frequency bands was done due to the analysis of the previous literature on the memory studies performed by using the EEG (Klimesch et al., 1999; Summerfield et al., 2005). In particular, a recent study of Werkle-Bergner and colleagues (Werkle-Bergner et al., 2006) reports as the increase of EEG activity in the theta and gamma bands for the encoding of complex material (visual as well as cognitive one) were particular relevant on the frontal and prefrontal hemispheres. Hence, beside the use of the z-score cortical power spectral maps, we used also the statistical analysis of these GFP indicators to derive sensitive indexes of the subject's appreciation and memorization of the commercial videos. Hence, the use of frontal leads for the estimation of the GFP indexes derives from the results reported in literature of the importance of these cortical areas for the encoding of the proposed audiovisual material (Werkle-Bergner et al., 2006).

4.1.2.1 Indexes Employed

In order to investigate the cerebral activity elicited during the observation of videoclips and analyze its variation according to the exposition of the brand advertised, we employed a series of indexes defined on the basis of the two GFP signals calculated in the theta and extended beta frequency bands. For each single TV commercial and subject, from these two cerebral signals we extracted information regarding the total number of peaks occurred along the clip length.

In fact, we define a peak in the two z-scored GFP waveforms in theta and extended beta band a part of the GFP waveform in which the values exceeds the threshold of $z = 1.96$, associated with a $p<0.05$ (uncorrected for multiple comparisons). We then defined the following Percentage index for the i_{th}

commercial on the z-score GFP, by defining for each advertisement presented to the subject the following parameter:

$$P_i = \frac{N_p \in T}{N_p} \tag{30}$$

where N_p is the number of peaks. The period of interest, T, was either the period in which the brand was exposed overtly in the commercial (BRAND period) or the period of time in which it was not exposed (NO BRAND period). In such a case the Percentage index is defined as the ratio between the number of peaks occurring in a particular segment of interest of the commercial and the total number of peaks observed in the all length of the advertisements.

4.1.2.2 Statistical Analysis of the GFP

Statistical analysis of the values of the z-score for the GFP and the Percentage Index variables were performed by using the Analysis of Variance (ANOVA) with different main factors. The log transformation was used to stabilize the variance of each variable by using the formula (Zar, 2009):

$$X' = \log(1 + |X|) \cdot sign(X) \tag{31}$$

where X is the original variable, X' is the log-transformed variable, sign(X) returns the sign of the variable X.

The main factors of the different ANOVAs performed on transformed GSR, HR, LF/HF and GFP variables are: BAND with the levels (THETA, BETA), BRAND with the levels (BRAND, NO BRAND), REPORT with the levels (RMB, FRG) for the group of subjects that remembered the particular spot (RMB) or viceversa (FRG), and again REPORT with the levels (LIKE, DISLIKE) to categorize the report of the subjects about the pleasantness (LIKE) or unpleasantness (DISLIKE) of the spot. The Greenhouse & Geisser correction was used where necessary for the violation of the sphericity hypothesis (Zar, 2009). Post-hoc analysis with the Duncan's test at the 5% statistical significance level was also performed.

Figure 18 shows the typical responses of the z-score variable obtained by the Global Field Power (GFP) computed on the subset of frontal electrodes in the theta (left panel) and the extended beta (right panel) frequency bands for a representative subject during the observation of a particular commercial within the documentary. It is worth of note that the GFP in the theta and in the beta band presents different series of peaks, occurring at different frames of the proposed commercial video. These frames are reported at the top of the images and the arrows point to the particular GFP relative to the frame illustrated. The colored areas in pink depicts the time interval in which the particular brand is overtly presented in the spot. In the case of Figure 18 such brand corresponds to a particular and well-known brand of biscuits in Italy. The time scale is in seconds.

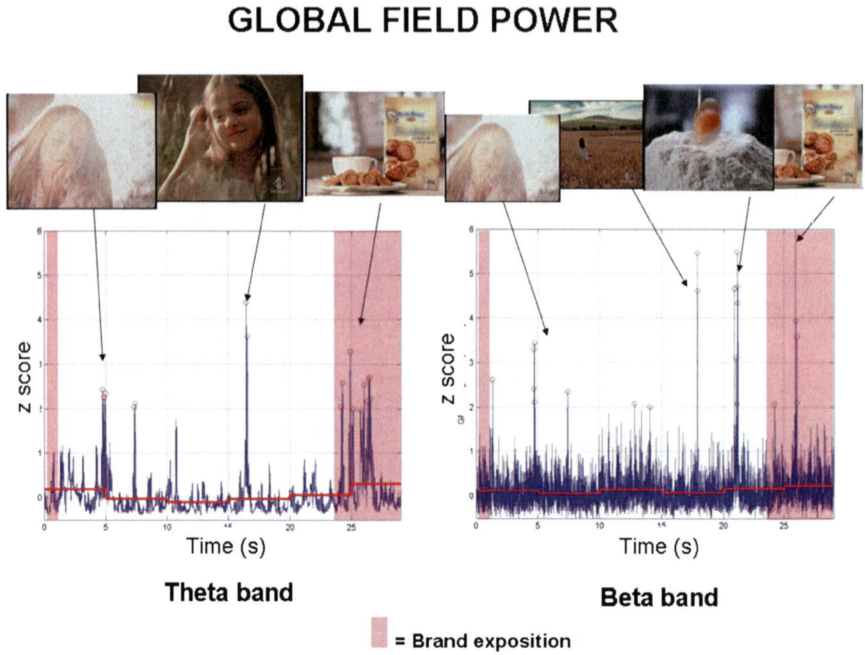

Fig. 18 Figure shows a typical responses of the z-score obtained by the Global Field Power (GFP) computed on the subset of frontal electrodes in the theta (left panel) and beta (right panel) frequency bands for a representative subject during the observation of a particular commercial within the documentary. The GFP in the theta and in the beta band presents different series of peaks, occurring at different frames of the proposed commercial video. These frames are reported at the top of the images and the arrows point to the particular GFP relative to the frame illustrated. The coloured areas in pink depicts the time interval in which the particular brand is overly presented in the spot. In the case of this Figure such brand corresponds to a particular and well know brand of biscuits in Italy. The time scale is in seconds.

The analysis of the variation of the average values of the GFP during the observation of the commercial videoclips will be performed by using three different factors for the ANOVA. The first main factor was the BAND one, with the levels theta and extended beta frequency (including beta and gamma conventional bands). The second factor is the BRAND one, with two levels (NO BRAND and BRAND) and the third factor is named REPORT, with the two couple of levels related to the memorization of the spot (RMB, FRG) or its pleasantness (LIKE, DISLIKE).

Hence, the ANOVA performed on the GFP index uses all these three main factors BAND, BRAND and REPORT. The ANOVA statistical outcome, in the case in which REPORT has the level RMB and FRG, is that no significant interactions occurred between all the employed factors. In particular, the BAND x BRAND x REPORT interaction is not significant at $p<0.73$, while also the other combination of factors are not significant, with BAND x REPORT with $p<0.78$ and BAND x BRAND with a $p<0.47$.

When the levels of the REPORT factor will became LIKE and DISLIKE then the ANOVA performed on the BAND and BRAND factors return a significant statistical interaction between all of them (BAND x BRAND x REPORT with a $p<0.038$). In addition, also the interaction between BAND x REPORT was statistically significant with a $p<0.048$, while the BRAND x REPORT interaction is not significant, having a $p<0.9$. The Figure 1 shows the BAND x REPORT interaction, with the representation of the average values of GFP in the two analyzed frequency bands (theta and extended beta bands). The average values are presented together their 2 standard deviation bars. The label GFP Theta and GFP Beta refers to the values of the GFP for the theta and beta band, respectively.

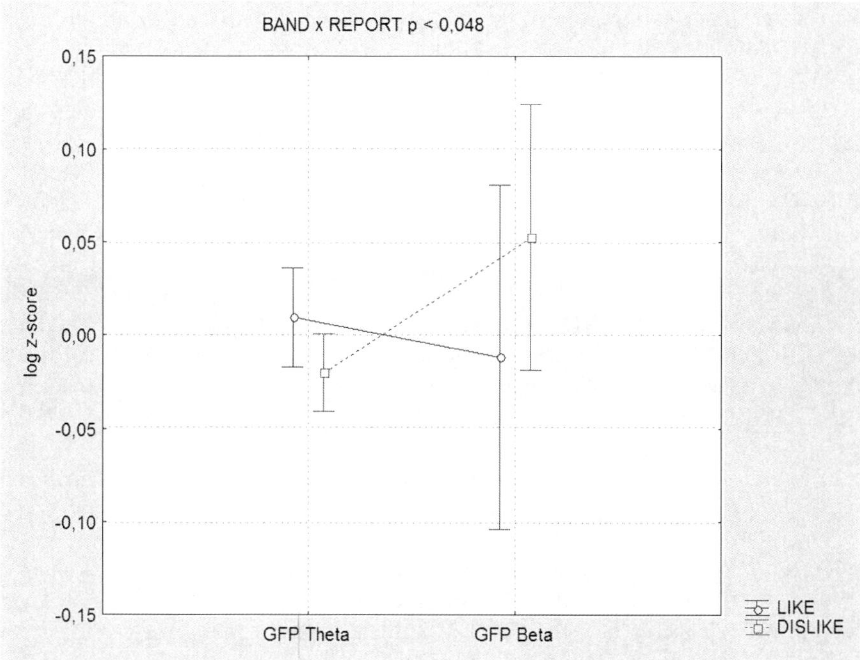

Fig. 19 Figure shows the BAND x REPORT interaction, with the representation of the average values of GFP in the two analyzed frequency bands (theta and beta). The average values are presented together their 2 standard deviation bars. The label GFP Theta and GFP Beta refers to the values of the GFP for the theta and beta band, respectively.

It is possible to appreciate the increase of the theta values for the GFP for the LIKE condition when compared to the DISLIKE one, while in the beta band this situation is reversed. Such data are congruent with the ones related to the cortical activity presented for the LIKE and DISLIKE groups in Figure 17 (upper left map).

Detailed post-hoc analysis with the Duncan procedure returned an increased statistically significant value of the GFP in the theta band in the LIKE condition ($p<0.05$) when compared to the DISLIKE group ($p<0.05$). It can be observed also a decrease of the beta and higher frequency band for LIKE group as compared to the DISLIKE one, but with no statistical significance ($p<0.4$).

The ANOVA reports a statistically significant interaction between the three factors BAND, BRAND and REPORT (with levels RMB/FRG) with a $p<0.026$. Another significant interaction exists between the factors BRAND and REPORT ($p<0.018$), as well as for BAND and BRAND ($p<0.001$). Duncan post-hoc tests reveals that the values of the Percentage index are higher for the theta band in the RMB group when compared to the FRG one, with a statistical significance of $p<0.0048$) during the BRAND level, and with a statistical significance of $p<0.0043$ in the case of NO BRAND level. No significant variation of the Percentage index is reported in the higher frequency bands and in the different BRAND levels.

In the case in which the REPORT levels are LIKE and DISLIKE, the performed ANOVA returned a statistically significant interaction of the main factors BAND and BRAND and REPORT with a $p<0.05$. Also the interactions between the couple of the main factors are statistically significant, i.e. BAND x BRAND ($p<0.001$), BAND x REPORT ($p<0.003$), BRAND x REPORT ($p<0.001$). The post hoc tests performed with the Duncan procedure show that also in this case the value of the Percentage index in the theta band is able to discriminate between the REPORT levels LIKE and DISLIKE during both the BRAND and NO BRAND time period of the spot ($p<0.0035$ and $p<0.045$, respectively). In addition, there is no statistical differences between the values of the Percentage index in the upper frequency band in order to discriminate the LIKE-DISLIKE conditions in the BRAND and NO BRAND periods ($p<0.92$ and $p<0.69$, respectively).

The use of the GFP index in the employed frequency bands has been analyzed with the ANOVA by taking its average values along the time period of the commercials affected by the BRAND exposition against the remaining time period (NO BRAND). The statistical analysis has returned a significant variation only for the LIKE/DISLIKE conditions, irrespective by the BRAND or NO BRAND period, with an increase of the GFP in the theta band for the LIKE condition. In addition, it was reported a decrease of the beta values for the LIKE group against the DISLIKE one. No significant variations of the GFP average values for the REMEMBER/FORGET levels. Also these particular results could be interpreted and linked to the previous reported studies, in which the increase of the power spectra in the theta band was already obtained (Aftnas et al., 2004; Astolfi et al., 2008a).

Beside the use of GFP with average values, the use of a particular Percentage index reporting the percentage of GFP peaks exceeding the statistical threshold in the BRAND and NO BRAND periods has been investigated. The results suggested that such Percentage index is sensitive to detect the differences between RMB and FRG in both the time period analyzed, as well as for the LIKE and DISLIKE conditions.

4.2 Patterns of Functional Connectivity

In a different study (Astolfi et al., 2008c) we also studied the cortical activity and connectivity occurring during the observation of TV commercials by the techniques of high resolution EEG and by the use of the functional connectivity estimates performed with the Partial Directed Coherence (PDC, (Baccalà and Sameshima, 2001; Astolfi et al., 2007a)). The extraction of significant descriptors of the estimated brain networks with the PDC was obtained with the use of particular graph theory indexes (De Vico Fallani et al., 2007). In order to estimate only the functional connectivity between cortical areas, a segmentation of fourteen Brodmann areas, thought to be of interest for this study, were considered as Regions of Interest (ROIs). Bilateral ROIs were the primary orbitofrontal and prefrontal areas, including the BA 8, 9, 10, as well as the Anterior Cingulate Cortex (ACC), the Cingulate Motor Area (CMA) and the parietal areas (BA 40, 5, 7). The labels of the cortical areas also have a postfix characterizing the considered hemisphere (R, right; L, left).

The analysis of 'where' the differences between the analyzed conditions occurred in the brain performed by the statistical mapping of power spectra was corroborated by the investigation on 'how' the different cortical areas are interconnected with the use of Partial Directed Coherence (PDC). In order to achieve our purpose, we analyzed the changes of incoming and outgoing flow for each ROI, according to the connection of the Granger's causality, by means of tools employed in the graph theory (De Vico Fallani et al., 2007). In fact, it is well known that a connectivity pattern can be treated as a weighted graph, where the nodes are the ROIs considered and the weighted arcs between nodes are the estimated connections between ROIs obtained by applying the PDC on the cortical data. It is then possible to apply tools already validated and derived from the graph theory to the estimated connectivity graphs during the task performance. In the following, the graph is described by N nodes (equal to the number of the ROIs considered here), and each arc of the graph from the i-th node toward the j-th node will be labelled with the intensity of the PDC value and will be described as wij. The NxN matrix of the weights between all the nodes of the graph is the connection matrix W. In particular, we would like to use the indices related to the strength of the estimated functional links between the cortical areas to characterize the behaviour of the estimated network during the visualization of the spot. Such indices will be described in the following. The simpler attribute for a graph's node is its degree of connectivity, which is the total number of connections with other points. In a weighted graph, the natural generalization of the degree of a node i is the node strength or node weight. This quantity has to be split into in-strength S-in and out-strength S-out indices, when directed relationships are being considered, as in the present case with the use of the PDC values.

These indexes are define in the following:

$$S_{in}(i) = \sum_j w_{ij} \tag{32}$$

where j = 1, 2, … N.

$S_{in}(i)$ represents, then, the amount of all the incoming arcs from the graph toward the node i-th, and it is a measure of the inflow of the graph toward such a node. A similar measure can be derived for the outflow from the i-th node of the graph, according to the following formula where the same conventions for (32) yields:

$$S_{in}(i) = \sum_j w_{ij} \tag{33}$$

Note that in this case the sum is upon all the outgoing weighted arcs that move from the i-th node towards all the other nodes of the graph.

The analysis of the strength indices has been addressed by means of a statistical procedure. Separate ANOVAs for each frequency band have been performed for the in-strength S_{in} and out-strength S_{out} with a significance level at 5%. In particular, the main factors of the ANOVAs are the within factor TASK with two levels, remembered (RMB) and forgotten (FRG), and the within factor ROI related to the ROIs employed in this study. Such ROIs include the BA 8,9,10, 5,7,40 of both hemispheres along with the cingulated motor area (CMA) and the anterior cingulated cortex (ACC). The Greenhouse and Geisser correction was used for the violation of the sphericity hypothesis (Zar, 2009). Post-hoc analysis with Duncan's test at the 5% statistical significance level was also performed.

The flow analyses related to the observation of commercial advertisements have been performed by means of the PDC values and some indexes of the graph theory according to the methods described in the previous section. For that purpose we analyzed the differences in terms of incoming and outgoing flows of Granger causality connections for each Region of Interest (ROI) analyzed, i.e. the values of S-in and S-out indices previously defined. These indices were computed by use of the values of the statistically significant functional connectivity estimates by PDC, performed at $p<0.05$, Bonferroni-corrected for multiple comparisons. Details of the estimation procedure for the computation of significant links between cortical areas by PDC are given in a previous publication (Astolfi et al., 2007a). The values of S-in and S-out indices were then computed for a selection of the ROIs employed in this study. In particular, the bilateral prefrontal areas (including the BA 8, 9, 10) and the parietal ones (including the BA 40, 5, and 7), with the Anterior Cingulate Cortex (ACC) and the Cingulate Motor Area (CMA).

The analysis of the incoming flow toward each particular ROI described by the S_{in} index was performed by use of the ANOVA with the main factors ROI and TASK. The test returned statistically-significant differences for the main factors ROI and TASK as well as for their interaction (ROI x TASK). The values of the statistical significance for the main factors ROI and TASK were always less than 5% ($p<0.05$) in all the four frequency bands analyzed. In particular, for the ROI x TASK condition we have $p<0.027$ in the Theta band, $p<0.023$ in the Alpha band, $p<0.032$ in the Beta band and $p<0.013$ in the Gamma band.

Figure 20 shows the average and standard deviation values of the S_{in} indices computed in the analyzed population for the two datasets computed for each ROI considered. All the four panels present in the Figure (A, B, C, D) present the average values of the S-in index for all the ROIs considered in the different frequency bands. It is possible to see that there is a higher value for the S_{in} index in the RMB than in the FRG condition in the analyzed population. In addition, post hoc analysis performed with Duncan's test at $p<0.05$ reveals a statistical increase of S_{in} values over the parietal areas of both hemispheres for the BA 7 bilaterally in all the frequency bands. In addition, in Theta, Beta and Gamma bands the parietal areas represented by the BA 40 and BA 5 results are statistically-significant bilaterally for the outflow during the RMB condition compared with the FRG condition.

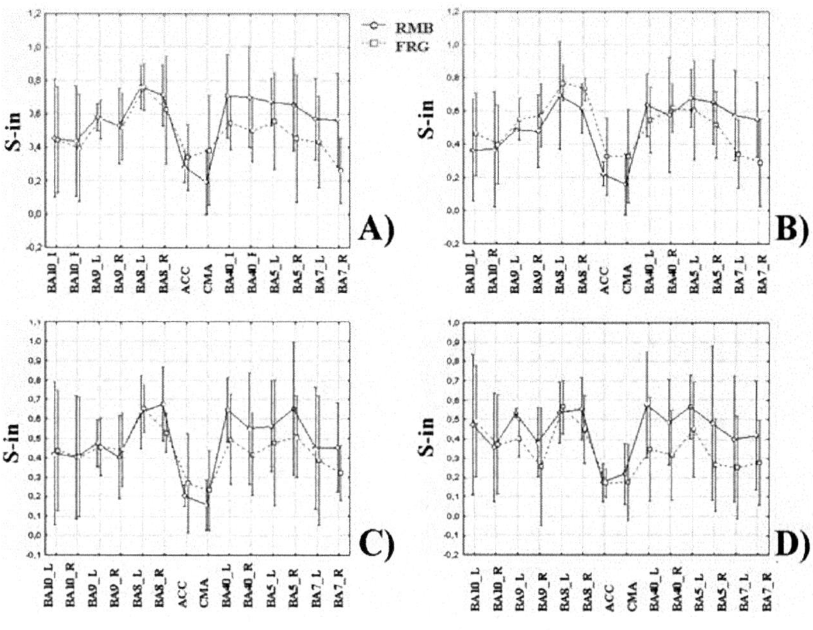

Fig. 20 Figure shows the average and standard deviation values of the S_{in} indices computed in the analyzed population for the two datasets (RMB, circles and continuous line; FRG squares and dotted lines) for each ROI considered. The four panels present in the Figure (A, B, C, D) are relative to the values of the S_{in} index in the four frequency bands considered. In particular, panel A refers to the variation of the S_{in} index across the different ROIs for the Theta band, panel B refers to the Alpha band, panel C to the Beta band and panel D refers to the Gamma band.

The analysis of the outflow from the different cortical areas, described by the S_{out} index, was also performed by use of the ANOVA. The main factors TASK and ROI results are statistically significant in all the frequency bands analyzed (always under $p<0.05$). For all concerns the statistical interaction between the factors TASK and ROI the tests returned statistically-significant values in all the analyzed frequency bands. In particular, we have $p < 0.012$ for the TASK x ROI factors in the Theta band, $p<0.032$ for the Alpha band, $p<0.042$ for the Beta band and $p<0.016$ for the Gamma band. Despite this significance of the interaction, the Duncan's post hoc tests reveal consistent statistically significant differences between the values of the S-out index in all the frequency bands analyzed only for the ACC and CMA cortical areas, the S-out values being comparable across the ROIs. The computed values of the outflow in the ACC and CMA cortical areas in the RMB condition are significantly greater than the S_{out} index computed in the FRG condition (all the frequency bands are below $p<0.03$). It must be noted that although high standard deviations are observed in these Figures, the ANOVA returns statistically significant differences between the factors used in the analysis. This was because of the superior statistical power of the ANOVA compared with the univariate statistical test.

4.3 Recording and Analysis of Autonomic Activity

The autonomic activity, both the Galvanic Skin Response (GSR) and the Heart Rate (HR), has been recorded with the PSYCHOLAB VD13S system (SATEM, Italy) with a sampling rate of 10 Hz. Skin conductance was recorded by the constant voltage method (0.5 V). Ag-AgCl electrodes (8 mm diameter of active area) were attached to the palmar side of the middle phalanges of the second and third fingers of the participant's non dominant hand by means of a velcro fastener. The company also provided disposable Ag-AgCl electrodes to acquire the HR signal. Before applying the sensors to the subjects' skin, their surface has been cleaned following procedures and suggestions published in the international literature (Schmidt and Walach, 2000; Fowles et al 1981; Venables, 1991). GSR and HR signals have been continuously acquired for the entire duration of the movie and then filtered and segmented with in-house MATLAB software. For to the GSR signal, we employed a band pass filter with a low cut-off frequency of 0.2 Hz, in order to split the phasic component of the electrodermal activity from the tonic one, and a high cut-off frequency of 1 Hz to filter out noise and suppress artefacts caused by Ebbecke waves (Schmidt and Walach 2000; Boucsein, 1992). Instead, the HR signal has been band pass filtered (high pass = 0.02 Hz; low pass = 0.6 Hz) in order to analyse the only frequency components due to variations of the sympathetic and parasympathetic nervous system regardless the ones associated to thermoregulatory cycles (Bernston, et al. 1997, Mendez, et al. 2006).

As previously explained, besides the autonomic activity of the subjects during the observations of the videoclips, we employed a part of the documentary to estimate the mean and standard deviation of the electrodermal activity and the cardiac signal in order to compute the z-score variables. These variables have been computed for each TV spot and subject, and then used to form the experimental datasets previously described (RMB, FRG, LIKE, DISLIKE) in order to perform a statistical analysis across time and conditions. Moreover, from the HR signal, the power spectrum density (PSD) has been calculated according to the Welch method (Welch, 1967). In this way, we obtained a signal in the frequency domain for the commercials analysed and for all subjects recorded. In order to compare the activity recorded during the advertisements with the one of the documentary, the PSD has been also evaluated for the REST period. In particular, the whole interval has been spanned with a series of time window of equal size of thirty seconds, each overlapped with the previous one of the 90%. In order to have a larger number of samples on which evaluating the values of mean and standard deviation of the PSD signal associated to REST period, the PSD has been computed for each time window. Spectral components were identified and then assigned, on the basis of their frequency, to one of two bands: Low Frequency (LF), [0.04, 0.15] Hz; High Frequency (HF), [0.15, 0.6] Hz (Malik et al., 1996). These components were obtained in absolute values of power (ms^2) and then converted in a-dimensional values of Z score as previously described. The Very Low Frequency (VLF) band, located in the lowest part of the spectrum, has been excluded from the present analysis since it is physiologically connected with long-term regulation mechanisms (Bernston et al. 1997; Mendez et al. 2006), not of interest for our purpose. Several studies indicates that the LF band corresponds to baroflex control of the heart rate and reflects mixed sympathetic and parasympathetic modulation of Heart Rate Variability (HRV); instead, HF band corresponds to vagally mediated modulation of HRV associated with respiration (Bernston et al. 1997; Malik et al. 1996; Mendez et al. 2006; Kreibig et al., 2007). For this reason, some researchers (Malik et al., 1996) propose the ratio LF/HF as index of the balance between the sympathetic and vagal activity. This was the measure we adopted to inspect and verify a likely correlation between the activity of the sympathetic nervous system with the subjects' answers regarding the memorization and the pleasantness of the TV commercials presented.

4.3.1 Statistical Analysis of the Autonomic Variables

As already illustrated for the analysis of the GFP values, also the autonomic values of GSR, HR and the ratio LF/HF have been z-score transformed. Moreover, the repeated-measures ANOVA has been adopted to highlight statistical differences.

While the GFP is a measure relative to the cerebral activity during the observation of the commercial video in different frequency bands, the Figure 25 presents the z-score waveforms of Galvanic Skin Response (GSR) and Heart Rate (HR) for the same representative subject. It is worth of note that the conventions employed in this Figure are similar to those presented in the Figure 18 being the axes scales the same (i.e z-score and time in seconds). However, Figure 21 also presents the mean values of the GSR and the HR in time interval of 5 seconds, as a red line overlaying the z-score waveforms. The z-scores for the autonomic signals did not reached in this particular subject a statistical significance for all concerns the GSR, while the threshold for the statistical significance was reached in some time instants for the HR values.

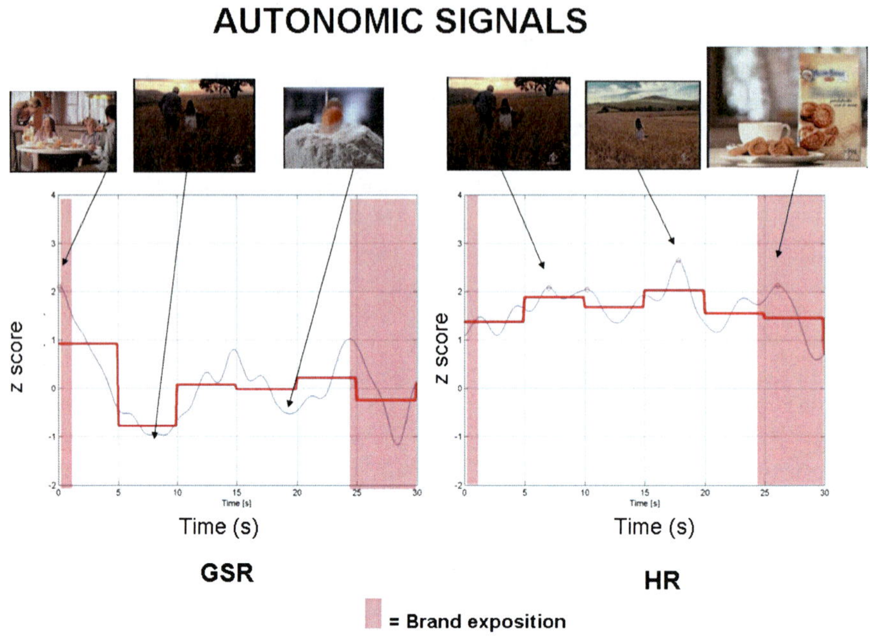

Fig. 21 Figure presents the z-score values for a typical subject during the observation of the same commercial video of Figure 21. Same conventions of the previous Figure are used. The mean values of the GSR and the HR are presented as the average on a time interval of 5 seconds, depicted in the panels with a red line overlaying the z-score waveforms.

The analysis of the average values of the autonomic variables gathered in the experimental group was performed by using another factor AUTONOMIC, with two levels (GSR and HR, including the transformed z-score of the GSR and HR recordings, respectively), and the already employed factors BRAND and REPORT with the same levels already described in the case of GFP.

As Figure 22 shows, the ANOVA performed in the case in which the levels of the factor REPORT are RMB and FRG returned a statistically significant interactions between the AUTONOMIC and REPORT factors ($p<0.023$), while the three way interaction between all the employed factors is not significant ($p<0.5$), as well as the interactions of AUTONOMIC x BRAND ($p<0.44$) and BRAND x REPORT ($p<0.63$).

Duncan post-hoc tests suggest that the values of the HR index are significantly increased during the observation of commercials in the RMB group when compared to the values of HR during the observation of commercials that will be forgotten (FRG group, $p<0.010$). On the contrary, the values of the GSR do not change significantly between the RMB and the FRG groups ($p<0.5$).

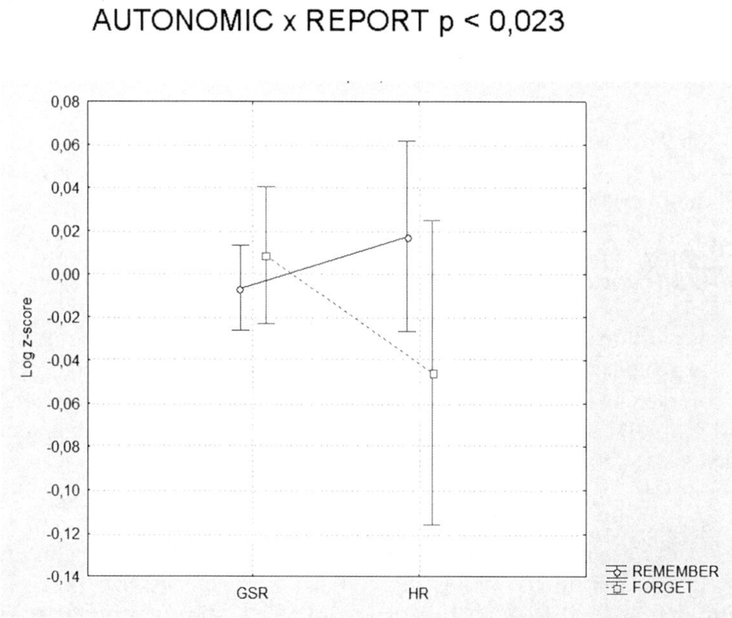

Fig. 22 Figure shows the AUTONOMIC x REPORT interaction, with the representation of the average values of GSR and HR for the two conditions (RMB, FRG) together with the representation of their 2 standard deviations by bars around the average values during the BRAND condition.

In the case in which the REPORT factor presents the levels LIKE/DISLIKE, the ANOVA returns a statistically significant interaction between all the factors employed, with a AUTONOMIC x BRAND x REPORT significance of $p<0.05$. The interactions between the main factors AUTONOMIC x REPORT is also statistically significant, with a $p<0.047$ while there are no interaction between the factors BRAND x REPORT ($p<0.48$). The post hoc analysis performed with the Duncan procedure returns that the values of the HR are statistically significantly higher in the LIKE group versus DISLIKE group in the BRAND condition ($p<0.048$) while there is a trend that is not statistically significant in the NO BRAND condition ($p<0.08$). There are no significance differences between the values of the GSR variable between the LIKE and DISLIKE groups, for both BRAND ($p<0.54$) and NO BRAND conditions ($p<0.43$).

4.3.1.1 Analysis of the Heart Variability Ratio

As indicated above, the LF/HF index has been used instead of the direct use of the LF and HF variables since in literature it has been suggested that such ratio is an indicator of the shift of the sympatho/vagal balance of the heart activity (Malik et al., 1996). In fact, it was important for the aim of this study to get an indication of the hypothesized shift of such balance during the exposition to commercial videos that will be memorized or judged pleasant when compared to the others.

A one-way ANOVA has been performed for the analysis of the index LF/HF derived from the heart rate variability, by using the factor REPORT with all the four levels included (i.e. RMB, FRG, LIKE, DISLIKE).

The results of the ANOVA reported a statistically significant increase of the LF/HF ratio for the LIKE group when compared to the DISLIKE group, with a statistical significance of $p<0.04$. In addition, the ratio LF/HF is also lower in the RMB group when compared to the FRG one ($p<0.042$).

4.3.2 Results from the Autonomic Indexes

The measurements of the heart rate and the heart rate variability (through the ratio LF/HF) reports a statistically significant difference when the experimental group are viewing commercial videos that resulted memorized (RMB vs FRG) as well as pleasant (LIKE vs DISLIKE) for the population analyzed. In particular, data suggested that heart rate variability index LF/HF is sensitive to the LIKE and RMB conditions being a greater value when compared to the DISLIKE and FRG conditions, respectively. This could be congruent with the general activation of the sympathetic system occurring during the observation of pleasantness images and videos. In addition, during the observation of the commercials for the LIKE condition, also the z-scored HR variables appear to be statistically different when compared to the DISLIKE group.

The results, offered by the analysis of the experimental data here provided, stated that the z-score levels of the GSR variable during the LIKE/DISLIKE or RMB/FRG maintained statistically similar values. If we adopt the hypothesis that GSR measures not the level of pleasantness of a situation but rather the level of arousal of the subject in a particular situation, as suggested in the specialized literature by several Authors (Nagai et al., 2004; Critchley, 2002), we could conclude that the level of arousal between the population did not change across the entire set of the commercial videos presented, irrespective of the RMB, FRG, LIKE or DISLIKE conditions.

Taken together, the indications provided by the autonomic measurements in the analyzed population suggest that HR, LF/HF are variables able to track the occurrence of memorization and pleasantness of the commercial videos. In addition, the proposed commercials did not elicit particular changes in the arousal of the investigated population. This is important since it was previously known that participants react to the viewing of highly aversive films with heart rate deceleration and a marked electrodermal increase (Baldaro et al., 2001; Palomba et al., 2000; Steptoe and Wardle, 1988; Oliveri et al., 2003). In this particular case, due to the particular nature of the videoclips presented (commercial advertisements) such orienting and aversive reaction was not generated.

Chapter 5
Cerebral Indexes of the Experienced Pleasantness

Nowadays, researchers are attempting to investigate the signs of the brain activity correlated with an increase of emotional involvement during the observation of commercial advertisements (Langleben et al., 2009; Vecchiato et al., 2010b). In fact, indirect variables of emotional processing could be gathered by tracking variations of the activity of specific anatomical structures linked to the emotional processing activity in humans, such as the pre- and frontal cortex (PFC and FC respectively; (Davidson and Irwin, 1999)). The PFC region is structurally and functionally heterogeneous but its role in the generation of the emotions is well recognized (Davidson, 2002). EEG spectral power analyses indicate that the anterior cerebral hemispheres are differentially lateralized for approach and withdrawal motivational tendencies and emotions. Specifically, findings suggest that the left PFC is an important brain area in a widespread circuit that mediates appetitive approach, while the right PFC appears to form a major component of a neural circuit that instantiates defensive withdrawal (Davidson, 2004; Davidson, 2000).

In this scenario, the purpose of this Chapter is to investigate the modulation of the Power Spectral Density (PSD) of the EEG rhythms elicited in the FC and PFC during the observation of commercial advertisements. In particular, the aim is to analyse the EEG frontal asymmetrical activations occurring while subjects are watching emotional scenes of a TV advertisement. Moreover, in the following, a cerebral index strictly correlated with the degree of pleasantness perceived by the experimental subjects is defined (Vecchiato et al., 2011b).

5.1 Pleasantness and Frontal Asymmetry

Since a clear role of the frontal areas have been depicted for the phenomena we would like to investigate (Davidson and Irwin, 1999; Davidson, 2002; Davidson, 2004), we used the left and right frontal and prefrontal electrodes of the 10-20 International System to compute the following spectral analysis. In such a case, we considered the following couples of homologous channels: Fp2/Fp1, AF8/AF7, AF4/AF3, F8/F7, F6/F5, F4/F3, F2/F1.

To study the EEG frontal activity, we compared the LIKE activity against the DISLIKE one by evaluating the difference of their average spectral values as follows:

$$Z = Z_{like} - Z_{dislike} \tag{34}$$

This spectral index has been mapped onto a real scalp model in the two bands of interest. Moreover, in order to investigate the cerebral frontal asymmetry, for each couple of homologous channels, we calculated the following spectral imbalance:

G. Vecchiato et al.: *Neuroelectrical Brain Imaging Tools*, BIOSYSROB 3, pp. 63 – 67.
DOI: 10.1007/978-3-642-38064-8_5

$$Z_{IM} = Z_{right} - Z_{left} \tag{35}$$

This index has been employed to calculate the correlation coefficient between the pleasantness score and the neural activity, in the theta and alpha band for each couples of channels we analyzed. Finally, we adopted the student's t-test to compare the Z_{IM} index between the LIKE and DISLIKE condition by evaluating the corresponding indexes.

In Figure 23 the scalp distribution of the z-score values for the theta and alpha bands is presented. It is possible to observe in the z-score distribution for the theta band a major activation for the condition $Z_{dislike}$ at the electrodes F2 and AF8, roughly overlaying the right frontal cortex (FC) and of the prefrontal cortex (PFC), although an enhance of spectral activity is also present in the left hemisphere (site F3). However, it is possible to observe also as the EEG spectral power is increased at the Fp1 electrode, for the Z_{like} dataset. The right side of Figure 23 shows the differences in EEG spectral power in the alpha band, mainly located at the electrode F1 that roughly overcame the left FC. Conversely, differences in EEG spectral power are also visible in the right hemispheres, at scalp sites AF8 and AF4 roughly overlying the right PFC.

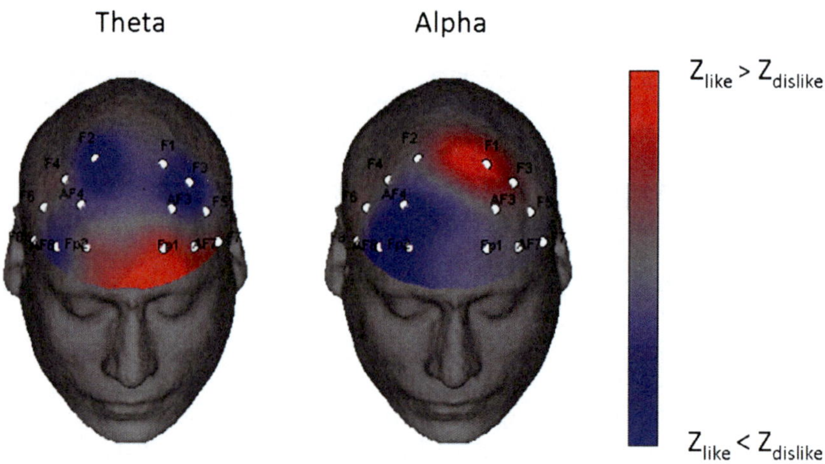

Fig. 23 The two scalp maps in Figure represent the Z x for the Theta (left) and Alpha band (right). Z values are mapped onto a realistic scalp model, seen from a frontal perspective. Colorbar codes scalp areas in which the LIKE spectral activity is greater than the DISLIKE (red) and regions in which the DISLIKE spectral activity is greater than the LIKE (blue). Grey indicates regions with no difference between the two experimental conditions.

In order to further investigate the frontal EEG asymmetry and its implications with the pleasantness our experimental subjects perceived, we calculated the correlation index between the imbalance index Z_{IM} described in Equation 35 and the pleasantness scores provided by the subjects, at each scalp location and frequency bands of interest. These results are summarized in Table 2. As to the theta band, we found out significant negative correlations at pre-frontal and lateral sites, while the only couple of electrodes F2-F1 present a significant positive correlation between the Z_{IM} and the pleasantness score. Instead, as far as concern the correlation in the alpha band, we obtained a significant positive correlation for the couple F2-F1.

Table 1. Correlation coefficients between Z_{IM} index and pleasantness score, for each couple of electrodes, for the theta and alpha band. Statistically significant values are highlighted in grey.

Correlation coefficients between Z_{IM} and pleasantness scores

	Fp_2-Fp_1	AF_8-AF_7	F_8-F_7	F_6-F_5	AF_4-AF_3	F_4-F_3	F_2-F_1
Theta	-0.17 (p=0.04)	-0.17 (p=0.04)	-0.21 (p=0.01)	-0.11 (p=0.16)	-0.02 (p=0.81)	0.07 (p=0.40)	0.17 (p=0.04)
Alpha	-0.11 (p=0.19)	-0.15 (p=0.07)	-0.15 (p=0.06)	-0.09 (p=0.26)	-0.02 (p=0.80)	0.05 (p=0.58)	0.16 (p=0.04)

Finally, in order to assess the different behaviour presented in Table 1, we performed a t-test analysis between the Z_{IM} values of the pre-frontal and lateral electrodes for the theta band (Fp2/Fp1, AF8/AF7, F8/F7, F6/F5) and the medial ones for the alpha band (AF4/AF3, F4/F3, F2/F1). The results of the statistical test, presented in Figure 24, revealed a significant difference of Z_{IM} values between the conditions LIKE and DISLIKE, in both theta ($t = -3.2$, $p = 0.0014$) and alpha band ($t = 2.2$, $p = 0.0298$). In fact, it is possible to observe a greater ZIM value for the DISLIKE condition in both theta and alpha bands.

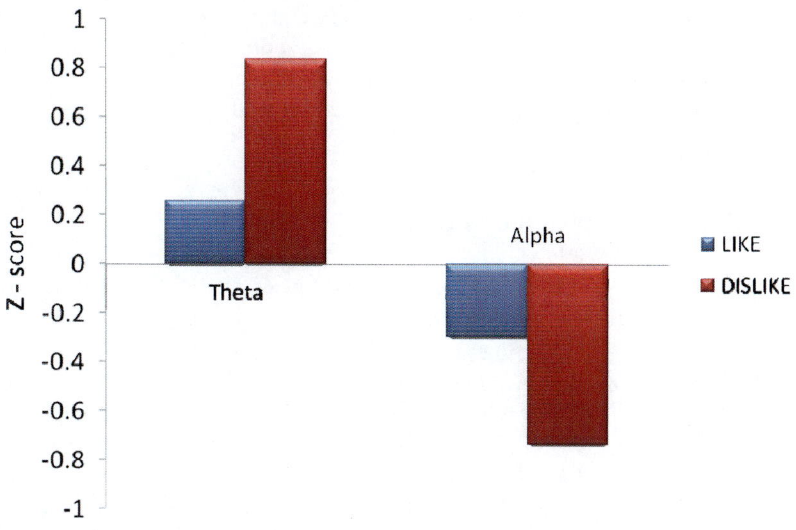

Fig. 24 Representation of the mean values of the Z_{IM} index for the LIKE (blue) and DISLIKE (red) conditions in the Theta (left) and Alpha band (right). Both differences are statistically significant (theta: $t = -3.2$, $p < 0.001$; alpha: $t = 2.2$, $p < 0.01$).

The results here presented showed a frontal asymmetrical activations of cerebral activity during the observation of pleasantness and not pleasantness commercial videos. Reported EEG power spectral maps distinguished the different activations between LIKE and DISLIKE conditions both in the theta and alpha band. It is worth of note that the most of activity in the left frontal hemisphere relates to the observation of commercials that have been judged pleasant by the analyzed population. On the other hand, the right frontal sites highlighted neuroelectrical activations concerning the observation of advertisements that have been judged less pleasant. Moreover, this imbalance in the activations was linearly correlated with the degree of pleasantness the subjects expressed. The correlation analysis revealed that pleasantness scores are significantly negatively correlated with the theta imbalance index, mostly concerning the pre-frontal and lateral frontal sites. Conversely, at the alpha frequencies the imbalance index is positively correlated around the medial frontal region. These data showed that the variations of the spectral index we defined are able to describe the degree of pleasantness feeling perceived by subjects while watching TV commercial ads. In particular, the scalp regions on the left frontal and pre-frontal areas are mostly activated when experimenting pleasant feelings. However, the right frontal lobe is more activated while watching commercials that have been judged unpleasant. Overall, the right frontal activity is significantly greater than the one in the left frontal lobe, both in theta and alpha bands. All together, these results are in line with previous findings suggesting the presence of an asymmetrical EEG activity when subjects experienced emotional stimuli (Davidson and Irwin, 1999; Davidson, 2002, 2004). Moreover, the greater spectral activity elicited in the right frontal areas during the observation of unpleasant TV ads could also be congruent with literature associating the insula/parainsula (Coan et al., 2006; Lamm et al., 2007) and the ventral anterior cingulated cortices (Somerville et al., 2006, Eisenberger et al., 2003) with the processing of negatively valenced emotions in social situations. Of course, these statements need to be confirmed by further studies employing the high resolution EEG techniques in order to estimate and investigate the relative cortical patterns.

5.2 Hedonic Evaluation of Logos

The research group of Handy and colleagues, instead, shifted their attention towards the rapid and emotional evaluation of advertising logos (Handy et al., 2010). Their study want to inspect whether the visuocortical processing of everyday images, like logos, can include an implicit hedonic analysis. In particular, they asked participants to identify, within a set of unfamiliar logos, those that were most liked or disliked. By means of an event related potentials (ERPs) analysis, they found out that visuocortical processing shows an increase of the early positive component (named P1), at central and parietal sites, along with an increase of the later negative component (named N2), at parietal and occipital sites, related to the observation of disliked logos. The idea at the base of this paper is to find electrophysiological signs correlated to the perception of liking or disliking particular advertising logos. Once this correlation was found, as expressed in the paper, it opens the way to an use of such ERPs P1 and N2 waves as a marker for the hedonic preferences of the consumer in front of the logos. This procedure could overcame the need to collect verbal preferences of consumers during the evaluation of different kind of logos, by replacing with an automatic and non-verbal evaluation of such hedonic preferences. Probably these information, related to N1 and P2, could be sufficient in the analysis of a simple logo while other type of variables, linked to more complex experiences by the users, could be employed to assess pleasantness of more complex marketing stimuli, such as an entire product (with the logo included). However, the possibility to have non biased clues about the "inner" perception of simple symbols in the brain of the consumers could have a marketing value and it will add a piece in the puzzle of the comprehension of the complex relationships between brand and consumers.

Chapter 6
Patterns of Cortical Activity Related to Public Service Announcements and Commercial Advertisements

The purpose of this Chapter is to illustrate the potential of the high resolution EEG techniques when applied to the analysis of brain activity related to the observation of TV commercials and Public Service Announcements (PSAs) to localize cerebral areas mostly involved. In particular, we want to describe how, by using appropriate statistical analysis, it is possible to recover significant information about cortical areas engaged by particular scenes inserted within the video clip analyzed. The brain activity was evaluated in frequency domain by solving the associate inverse problem of EEG with the use of realistic head models. Successively, the data analyzed were statistically treated by comparing their actual values to the average values estimated during the observation of the documentary.

By employing the same neuroelectrical dataset and methods of the experiment described in Chapter 5, we divided the physiological activity recorded into other four different experimental groups (Vecchiato et al., 2010c). The first pool was related to the activity collected during the viewing of the PSA (CAR) clips that the subjects had correctly remembered, and this dataset was named PSA_{RMB} (CAR_{RMB}). The second pool was related to the activity collected during the observation of the PSA (CAR) that had been forgotten by the subjects, and this set was named PSA_{FRG} (CAR_{FRG}). In such a way, the four different EEG datasets considered in this study are PSA_{RMB}, $PSA_{FRG,}$ $CAR_{RMB,}$ CAR_{FRG} which refer to our four different experimental conditions. Finally, the EEG activity elicited during the observation of the documentary was also analyzed and a final pool of data, related to this state, was generated with the name REST. This REST period was taken as the period in which the subject looked at the documentary. We took into account a two minutes long sequence of the documentary, immediately before the appearance of the first spot interruption, employed in order to minimize the variations of the spectral responses owing to fatigue or loss of concentration.

The Power Spectral Density has been calculated for all EEG segments, which are associated to the observation of the PSA and CARs advertisements, by means

G. Vecchiato et al.: *Neuroelectrical Brain Imaging Tools*, BIOSYSROB 3, pp. 69–74.
DOI: 10.1007/978-3-642-38064-8_6

of the mathematical procedures described in the previous chapter. For each subject, these PSDs have been evaluated for all the frequency bands of interest and then compared with the values of power spectral density elicited by the observation of the documentary (REST segment). The contrast between these two experimental conditions has been performed by estimating the z-score variables. In such a way, the cortical distributions of the z-scores related to the observation of the commercial videoclips have been organized into four different groups: the first two refer to the cerebral activity elicited by the observation of PSAs which have been remembered by the subjects during the interview (PSA_{RMB}) and those which have been forgotten during the same interview (PSA_{FRG}); in the same way, we grouped the cortical activity elicited by the observation of the CAR TV spot by separating that associated to the EEG segments of subjects who remembered such commercials (CAR_{RMB}) from that associated to the EEG segments of subjects who forgotten such commercials (CAR_{FRG}). Contrasts will be made among the cortical z-score distributions of the defined experimental groups by comparing the cerebral activities related to both the RMB and FRG conditions within the same category of commercials (PSA_{RMB} vs PSA_{FRG}; CAR_{RMB} vs CAR_{FRG}), and by comparing the same cerebral activities between the two categories of advertisements (PSA_{RMB} vs CAR_{RMB}; PSA_{FRG} vs CAR_{FRG}). The resulting statistical spectral maps of z-scores highlight the cortical areas in which the estimated power spectra statistically differ between the populations considered. The following maps refer to the contrast between the experimental populations in the theta (upper left), alpha (upper right), beta (lower left) and gamma (lower right) frequency bands.

The Role of Public Service Announcements

Local governs of European countries, and also across the world, are called to disseminate information about health risky habits, promoting instead healthy life style, in order to increase the health of their citizens. In this context Public Service Announcements (PSAs) are non-commercial broadcasted ads intended to modify public behavior. PSAs are at the core of many public health campaigns against smoke, fatty foods, abuse of alcohol and other possible threats for the health of citizens. But the content of these PSAs could be also directed for the promotion of "positive" social collective behavior, such for instance against racism, supporting the integration of different cultures in the country, or for instance promoting a healthy drive style, for the road security. When effective, PSAs carry a great public health benefit (Biener et al., 2000; Emery et al., 2005). However, the lack of reliable, quantitative and objective means of ad evaluation is one of the impediments to better PSA outcomes. In addition, not well designed PSAs are going to have counter effects with respect to their desired goals (Wakefield et al., 2003). On the other hand, commercial advertisings are announcements of material or goods of potential interest for the public, usually offered for the purchase. So while the two kind of announces are often similar for realization, they are different for the message convoyed to the public.

6.1 Statistical Maps of the Spectral Cortical Activity during the Observation of the PSAs

Figure 25 presents four cortical maps in which the brain is viewed from a frontal perspective. The color scale on the cortex coded the statistical significance: where

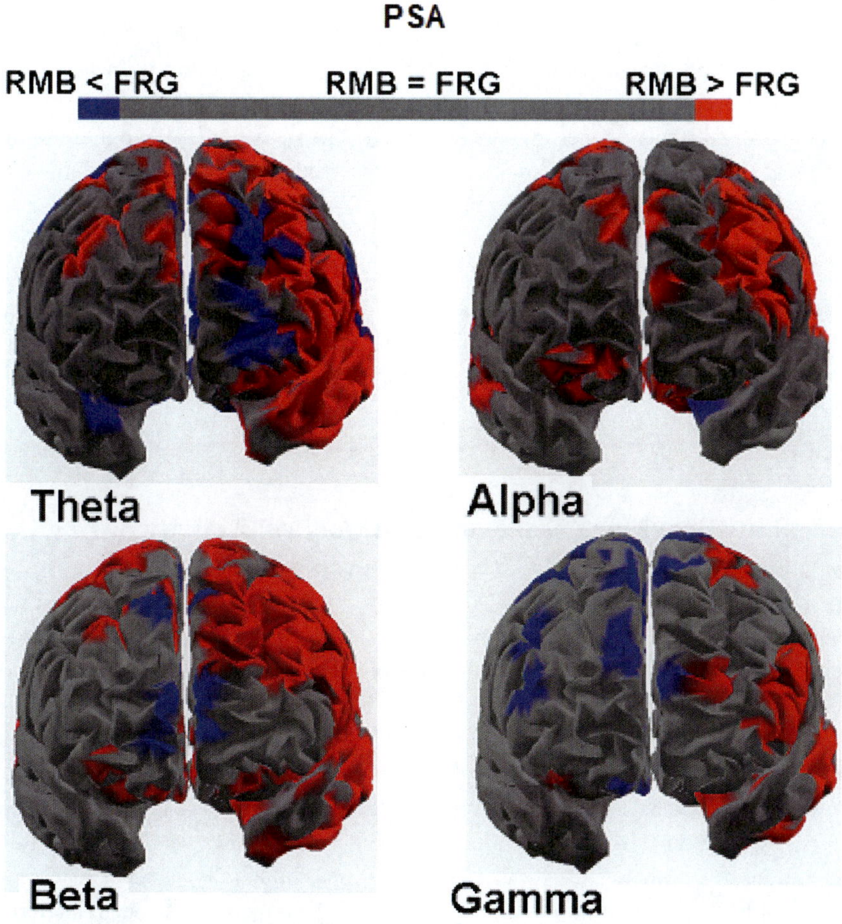

Fig. 25 Figure presents four cortical z-score maps, in the four frequency bands employed. Colour bar represents cortical areas in which increased statistically significant activity occurs in the PSA_{RMB} group (subjects watching PSAs that will be remembered) when compared to the PSA_{FRG} group (subjects watching PSAs that will be forgotten) in red, while blue is used otherwise ($p < 0.05$ Bonferroni-Holm corrected). Grey colour is used to map cortical areas where there are no significant differences between the cortical activity in the PSA_{RMB} and PSA_{FRG} groups.

there are cortical areas in which the power spectrum does not differ between the two populations, a grey color was employed. The red color was employed when the cortical areas present a statistically significance power spectral activity greater in the population that remembered the commercial videos (RMB) with respect to the other, while the blue color coded the opposite situation (i.e. the power spectral activity in the group that forget the commercial videos is greater with respect to the brain activity in the group that remembers the ads).

Figure 25 shows an increase of cortical activity in all the bands of interest that is prominent on the left frontal hemisphere for the PSA_{RMB} group. In particular, the cerebral activity in theta band presents a statistically significant activation among frontal areas of the left lobule for the PSA_{RMB} group, although there are also some prefrontal areas depicted by the activation due to the PSA_{FRG} population. This sensible activation for the PSA_{RMB} dataset is still clearer by examining the spectral cortical activity in the alpha and beta bands: the cortical maps relating to these frequency bands are characterized by a statistically significant PSD values located in a wide area of the left frontal hemisphere. Instead, the spectral map of the gamma band presents both significant activations of activity for the PSA_{RMB} group and for the PSA_{FRG} population in cortical regions located in the left and in the right hemisphere of the frontal lobule respectively.

6.2 Statistical Maps of the Spectral Cortical Activity during the Observation of the CAR Commercials

Figure 26 presents the contrast between the CAR_{RMB} and CAR_{FRG} groups in the four frequency bands considered in this analysis by using the same convention of Figure 30. The significant increase of the frontal activity in the theta band is clearly visible in the CAR_{RMB} group (in red) when compared to the CAR_{FRG} one (in blue), in the left upper part of the Figure 30.

Scattered increased of cortical activity on the left hemisphere is also present in the CAR_{FRG} group. Instead, the alpha frequency map shows scattered increased activations for both CAR_{RMB} and CAR_{FRG} groups, bilaterally located in the left and right hemispheres. Similar considerations can be done for the spectral activity in the beta band since it presents only a few spots of significant activations for the CAR_{RMB} population. The cortical activity in the gamma band is greater in the CAR_{RMB} group which shows a large statistically significant activations spreading both in the pre and frontal regions of the left hemisphere and in the frontal areas of the right hemisphere.

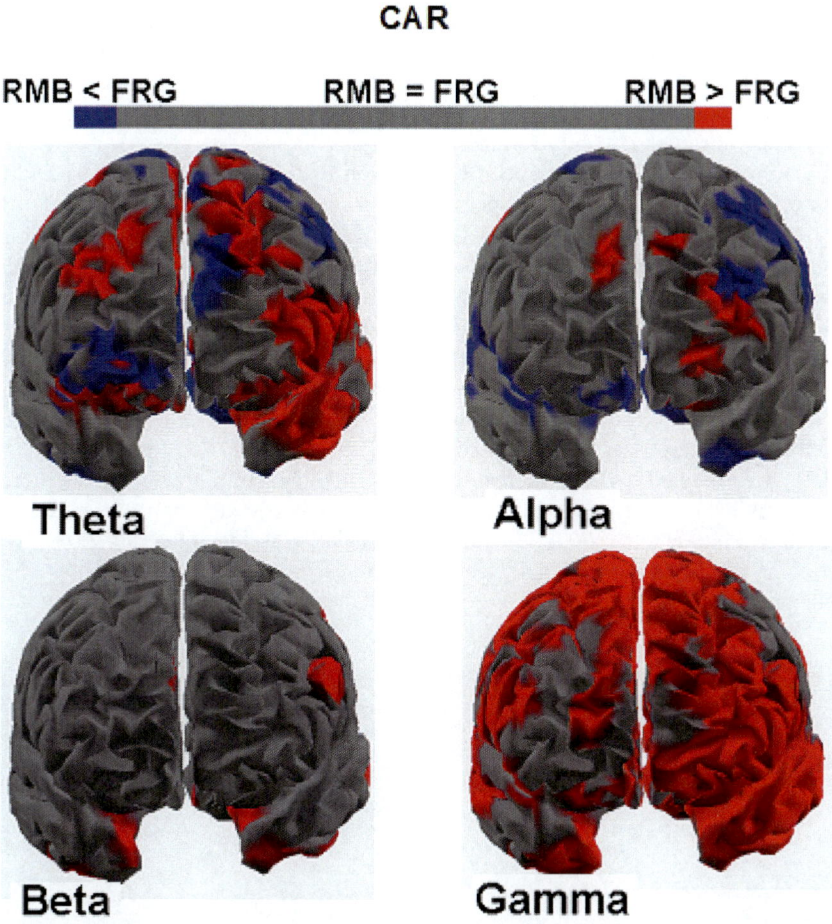

Fig. 26 Same conventions than in the previous Figure except the use of red colour to code the cortical areas in which the brain activity of the CAR_{RMB} group (subjects watching commercials about cars that will be remembered) is significantly higher than the activity of the CAR_{FRG} group (subjects watching commercials about cars that will be forgotten). The blue colour is used to map brain areas in which the activity of the CAR_{FRG} group is significantly higher than the activity of the CAR_{RMB} group.

6.3 Analogies and Differences in the Patterns of Activations

Taken together, the results showed in this Chapter indicate the cortical activity in the theta band on the left frontal areas was increased during the memorization of commercials. Such results are then congruent with the role that has been advocated for the left pre and frontal regions during the transfer of sensory percepts from the short term memory toward the long-term memory storage by the

HERA model (Summerfield and Mangels, 2005; Werkle-Bergner et al., 2006; Astolfi et al., 2008a; Tulving et al., 1994). In fact, in such model, the left hemisphere plays a key role during the encoding phase of information from the short term memory to the long term memory, whereas the right hemisphere plays a role in the retrieval of such information.

Moreover, apart from the specific activation of the prefrontal left cortices (as expressed in the increase of the spectral power in the theta and gamma bands) we also found out an enhance of activity in different part of the cortex in the alpha and beta bands among the population who remembered the PSAs we proposed. This result is not evident by inspecting the cortical maps associated to the CARs. This fact could be interpreted by assuming the existence of different cortical areas processing these two different kinds of audiovisual stimuli. In fact, we could speculate that the fact that the PSAs are related to issues treating health and possible threats for the individual could elicit an increased level of attention when compared to the observation of standard CARs. Such increased level of attention could be reflected by the increase of the EEG spectral power in the upper frequency bands, often linked to the increase of aspecific attention to the external world (Oliveri et al., 2003; Urbano et al., 1998; Babiloni C et al 2003). The different cortical areas elicited seems hence to be a property of PSAs when compared to the CARs.

Chapter 7
Cultural Similarities and Differences in Advertising between Western and Eastern Consumers

Advertisings related to the fruition of carbonated beverages are intensively presented on the usual TV programs worldwide. Such as advertisements are also present on journals, street posters, fast foods and supermarkets in a pervasive way and could have a deep impact on human behavior. Recent functional neuroimaging studies have begun to investigate how commercial brand information is processed in the brain (Frank, 2003; McClure et al., 2004; Deppe et al., 2005). Although the experimental designs vary, each of these studies report activity in ventral and/or medial prefrontal cortex during the contemplation or consumption of familiar brand-name products. Since lesion studies indicate ventromedial prefrontal cortex (VMPC) is critically involved in emotion, emotional regulation and decision-making (Koenigs et al., 2007), ventromedial prefrontal activations can be interpreted as evidence for emotion playing a pivotal role in brand preference (Frank, 2003; Deppe et al., 2005). Psychological and sociological studies documenting the use of emotional appeal in advertisements further support the significance of emotion in brand preference formation (Shadel et al., 2002; Anderson et al., 2005). Whereas the role of prefrontal cortices is highlighted in the generation of appreciation for a brand, it is not really addressed the issue of how this appreciation is spread across different cultural models, i.e. across Western and Oriental people. In fact, it is well known like different cultural model in Western and Oriental culture leads to different appraisal of the same experience or situation. Hence, it is of value to understand if, from a cerebral point of view, people educated in different culture could similarly react to the same kind of advertisement related to carbonated drinks.

The aim of the present Chapter is to illustrate analogies and differences in cerebral activations between a Western and an Eastern population during the fruition of TV commercials (Vecchiato et al., 2012b, 2012c). In particular, we show that the related EEG memorization patterns highlight a common behavior between the two populations, although the emotional activity illustrate that Western and Eastern could differently react to the observation of the same commercial. This finding could be helpful to marketers to understand if it is needed to realize different commercials, for the same product, to be broadcasted in culturally different countries.

G. Vecchiato et al.: *Neuroelectrical Brain Imaging Tools*, BIOSYSROB 3, pp. 75–84.
DOI: 10.1007/978-3-642-38064-8_7

7.1 Increase of Theta EEG Activity Correlates with Memorization of Advertisements for Chinese and Italian Subjects

In this section we report the results related to the cerebral activity gathered during the fruition of similar advertisements related to very popular carbonated beverages (Coca Cola and Pepsi Cola) in group of Western and Oriental people (Vecchiato et al., 2012b). In particular, we conducted two series of experiments, both in Italy and in China aimed at collecting EEG data related to the fruition of Coca Cola and Pepsi Cola advertisings. We used EEG technologies in order to exploit the possibility of the provided time resolution, on a milliseconds base, useful to track the changes in the cerebral activity related to the succession of scenes in the TV commercial. Analysis was focused on the activity of the prefrontal cortices of such videoclips in both Western and Oriental populations.

Since for the phenomena we would like to investigate a clear role of the frontal areas in the theta band have been depicted (Summerfield and Mangels, 2005), we used frontal electrodes to compute the normalized GFP indexes and employed the signals in the theta band to define a Memorization Index (Vecchiato et al., 2010b).

The filtered theta trace has been employed to calculate the Global Field Power (GFP) for each segment. Afterwards, they all have been averaged and normalized by calculating the z-score index and considering the REST dataset as baseline. The formula employed is presented in the following:

$$GFP_Z(t) = \frac{GFP_{SPOT}(t) - mean(GFP_{REST}(t))}{std(GFP_{REST}(t))} \tag{36}$$

The filtered EEG signals were subjected to the computation of the Global Field Power by taking into account the signals gathered from the following frontal and prefrontal electrodes of the 10-10 International System: AF3, Fp1, F5, AF4, Fp2, F6. In the following, we contrast the z-score activity related to the RMB population against the one belonging to the FRG group by subtraction. The z-score comparison has been performed on GFP waveform and on the averaged GFP calculated in particular scenes of interest for both commercials. Statistically significant results were obtained when z-score exceeds 3.0 in absolute value the (corrected z-score for multiple comparisons).

In the following Figures we show time-varying changes of the GFP as z-score variable in the theta band during the observation of both Italian (Figure 27 and Figure 28) and Chinese (Figure 29 and Figure 30) TV commercials of two different carbonated beverages.

In Figure 27 and Figure 28 it is possible to observe the variations of the indexes of memorization regarding the Italian subjects, during the observation of the commercial. The videoclip is related to a series of short stories in which the presence of Coca Cola makes happy the actors. The sequences usually present a single person performing jokes to other people while he/she's drinking Coca Cola. The ad closes with the presentation of the brand and product. In Figure 27, as far as concern the difference of GFP waveform between the RMB and FRG groups, the cerebral variable related to the memorization process presents three peaks of activation. In particular, the first film segment of interest is at the beginning of the commercial, 4'', when a girl is making a joke. The second enhance of activity occurs during the observation of funny scenes around the 9''.

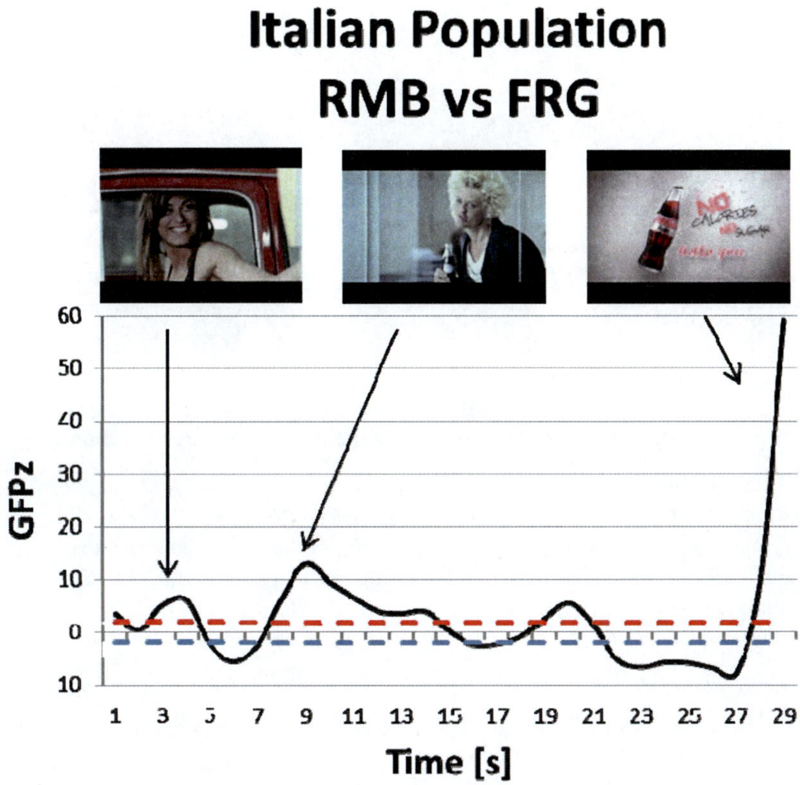

Fig. 27 Representation of the GFPz (black line) as difference between GFP_{RMB} and GFP_{FRG} in the theta band for the Italian population during the observation of the TV commercial advertising a carbonated beverage. On the x-axis, the duration of the spot; on the y-axis, the z-score values. Dotted lines represent the statistic threshold for significant increase of activity for RMB group (red) and FRG group (blue), $p<0.05$. Film segments are showed when increase of cerebral activity occurred in the RMB population with respect to the FRG one.

Finally, the last peak of activation related to the RMB dataset is associated to the exposition of the brand and the product advertised along with the payoff, located between the 28" and the 30".

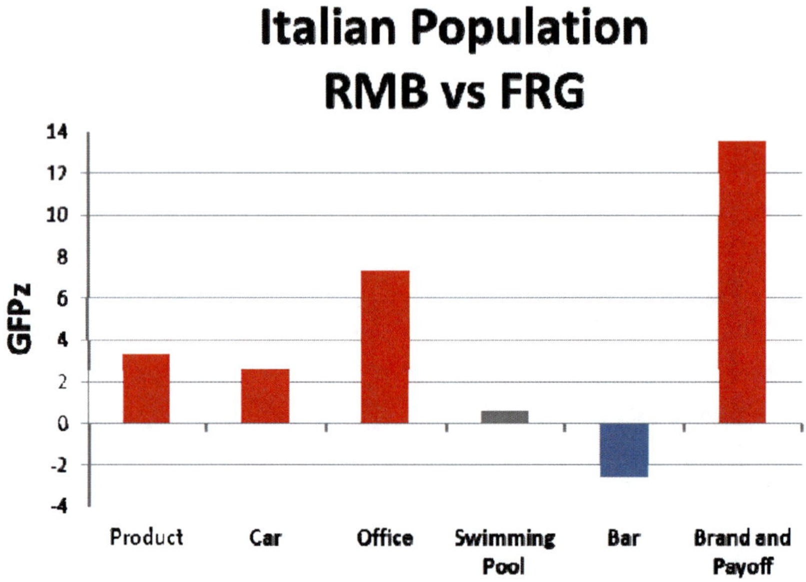

Fig. 28 Representation of the GFPz as difference between GFP_{RMB} and GFP_{FRG} in the theta band for the Italian population during the observation of the TV commercial advertising a carbonated beverage. Comparisons refers to main scenes of interest. Red (blue) color represents a significant increase of activity for RMB (FRG) group, $p<0.05$. Gray color means no statistical difference.

In Figure 28 we compare the z-score theta activity, between the RMB and FRG group, in main scenes of the Italian commercial. Differences highlight a statistically significant increase ($z > 3.5$) of activity for the RMB group in most of the scenes, in particular during the payoff and brand exposition.

In Figure 29 and Figure 30 we show the results obtained by analyzing the cerebral activity of the Chinese dataset. The video presented is related to a young male using the Pepsi can as a microphone. Successively, when he sings, all different persons around a city listen to him through the same Pepsi can. There is an alternation of scenes in which the young male sing alone and groups of persons listen to and sing by using the same Pepsi can. The ad closes with the presentation of a series of Pepsi products.

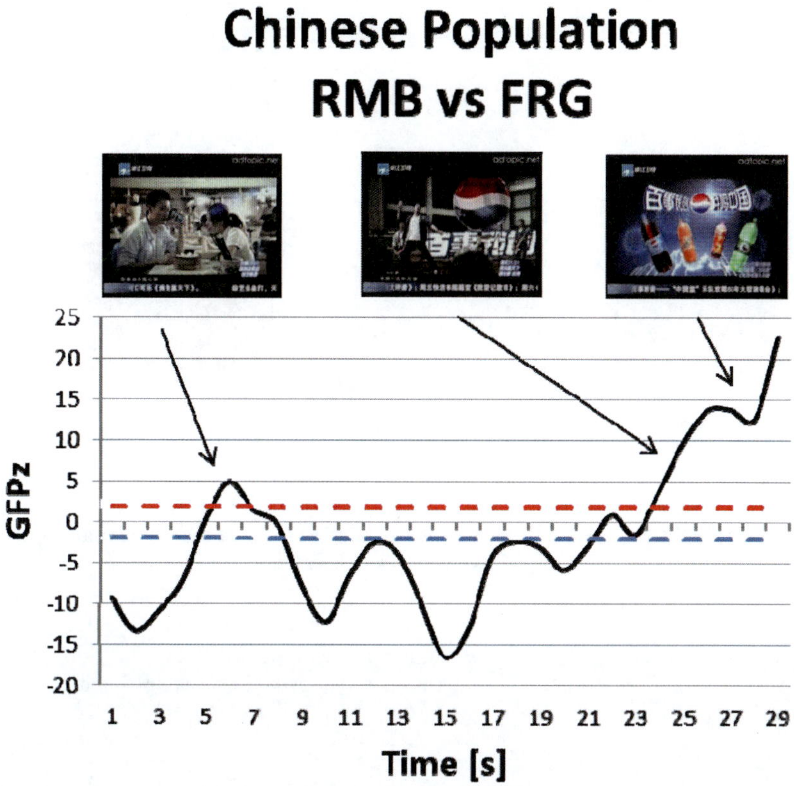

Fig. 29 Representation of the GFPz (black line) as difference between GFP_{RMB} and GFP_{FRG} in the theta band for the Chinese population during the observation of the TV commercial advertising a carbonated beverage. Same layout of Figure 27.

In Figure 29, as far as concern the difference of GFP waveform between the RMB and FRG groups, the memorization index mainly presents three increases of activity. The first one is located around the 6" and it is associated to the observation of people using a can as earphone. The second peak of activation is located between the 25" and the end of the commercial. It is related to the observation of the singer on the stage, showing the logo, and to the final payoff. All these peaks are statistically significant.

In Figure 30 we compare the theta activity, between the RMB and FRG group, in main scenes of the Chinese commercial. Differences highlight a significant increase of activity for the RMB group in most of the scene, in particular during the logo and brand exposition.

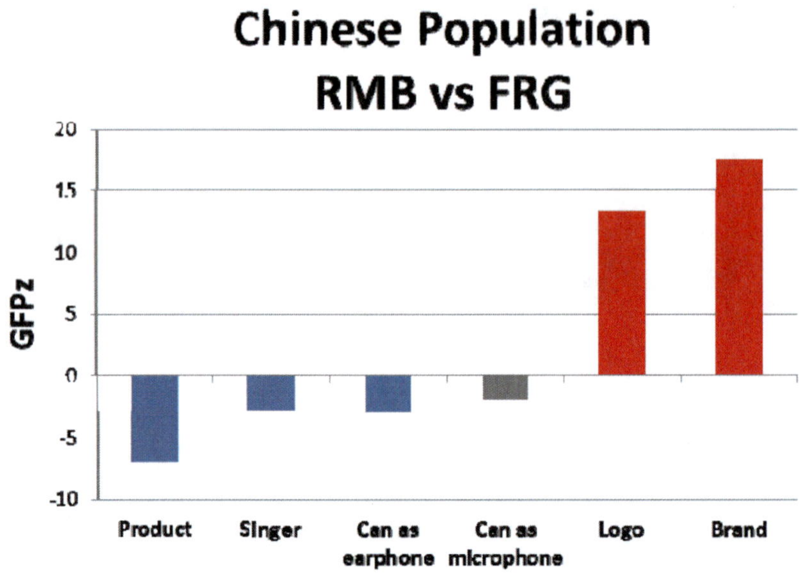

Fig. 30 Representation of the GFPz as difference between GFP_{RMB} and GFP_{FRG} in the theta band for the Chinese population during the observation of the TV commercial advertising a carbonated beverage. Comparisons refers to main scenes of interest. Same layout as Figure 28.

These results show how the enhance of the GFP calculated on left frontal electrodes in the theta band for the RMB population is related to the observation of particular frame segments presenting the product advertised, in particular the payoff and brand exposition, for both TV commercials analyzed.

As shown in the previous Figures, peaks of frontal theta activity emerge while subjects are watching the brand and the beverage advertised and particular funny scenes. This phenomenon is reasonably connected to the memorization process of the product shown since it has been successfully remembered in the later interview. Moreover, the biggest increase of activity related to the memorization of both commercials mainly occur during the brand exposition.

7.2 Differences in Emotional Processes of TV Commercials in a Group of Eastern Subjects

In this section, our purpose is to illustrate the modulation of the Global Field Power (GFP) of the EEG rhythms elicited in the frontal and prefrontal cortex during the observation of commercial advertisements. In particular, the aim of the present study is to analyze the level of memorization, attention and emotion perceived while Eastern subjects are watching an Eastern and a Western version of the same TV commercial (Vecchiato et al., 2012c). Moreover, by monitoring autonomic activity such as the heart rate (HR) it is possible to assess the emotional state of the subject. In fact, the link between the heart rate and emotions has been already suggested (Montano et al., 2009).

Each EEG trace has been band pass filtered in order to isolate the spectral components in the theta, lower alpha and upper alpha bands from the whole EEG spectrum.

The filtered traces have been employed to calculate the Global Field Power (GFP; Lehmann and Skrandies, 1980). Since for the phenomena we would like to investigate a clear role of the frontal areas have been depicted (Davidson, 2004; Klimesch, 1999; Werkle-Bergner et al., 2006) we used the frontal electrodes to compute the GFP indexes used in the following of this study. In order to summarize the properties of the cerebral activation for the analysed ads we used the theta, lower alpha and upper alpha bands to define the Memorization, Attention and Pleasantness indexes, respectively (Vecchiato et al., 2010b, 2011; Davidson, 2004; Klimesch, 1999; Werkle-Bergner et al., 2006;). The filtered EEG traces were subjected to the computation of the GFP by taking into account the signals coming from the following frontal and prefrontal electrodes of the 10-10 International System: AF3, F3 to compute the Memorization Index (MI); Fpz, AF3, F3, AF4, F4, Fz to calculate the Attention Index (AI); homologous channels AF3, AF4, F3, F4 to evaluate the Pleasantness Index (PI). As to the Attention Index, we reversed the GFP waveform in order to have the activity of de-synchronization pointing up. As far as concern the Pleasantness Index, it has been defined by taking into account the frontal EEG asymmetry's theory by Davidson (2004) as already investigated in a previous study (Vecchiato et al., 2011). Hence, the formula defining the PI is the following:

$$PI = GFP_{right} - GFP_{left} \tag{37}$$

Where the GFP_{right} and GFP_{left} stand for the GFP calculated among right (AF4, F4) and left (AF3, F3) electrodes, respectively. The GFP signals of each subject have been averaged to obtain a mean waveform to be compared between the two commercials. Statistical analyses will be performed by using t-test at $p<0.05$ after the execution of the Bonferroni correction for multiple test.

In the following we are going to illustrate the results we achieved in the experiment described above. The following Figures show the values of MI, AI, PI and HR in the seven segments of interest by comparing the Western and the Eastern version of the analyzed TV commercials.

Figure 31 presents the GFP values for the Memorization and the Attention Index for both advertisements. The seven columns show the difference of the cerebral variables existing between the two TV commercials. For both Memorization and Attention Index it is possible to observe that each segment of interest is characterized by the same average value of GFP. Overall, this result suggest that from a cognitive point of view the two version of the advertisement elicit the same level of memorization and attention in the population analyzed.

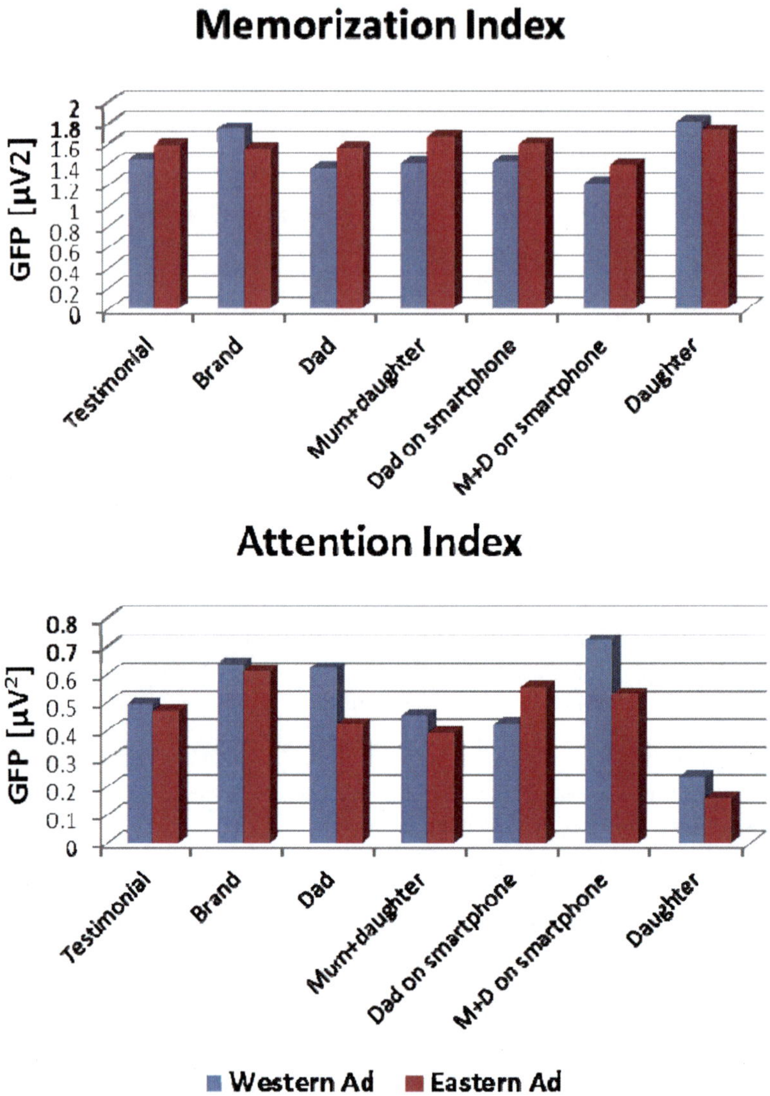

Fig. 31 Figure presents the average values of GFP in the theta (Memorization Index, upper row) and lower alpha band (Attention Index, lower row), respectively. Each column refers to a single segment of interest showing the values of the cerebral variables for both Western (blue) and Eastern (red) advertisement. No statistical difference among the experimental conditions.

Figure 32 presents the cerebral values for the Pleasantness Index (PI) and Heart Rate measurements for both advertisements. The seven columns show the difference of the cerebral and autonomic variable existing between the two TV commercials. As far as concern the Pleasantness Index, we observe that the

Western Ad is characterized by negative GFP values for all segments of interest. According to the definition of the PI, such result means that the whole commercial is mostly perceived as unpleasant for the population analyzed.

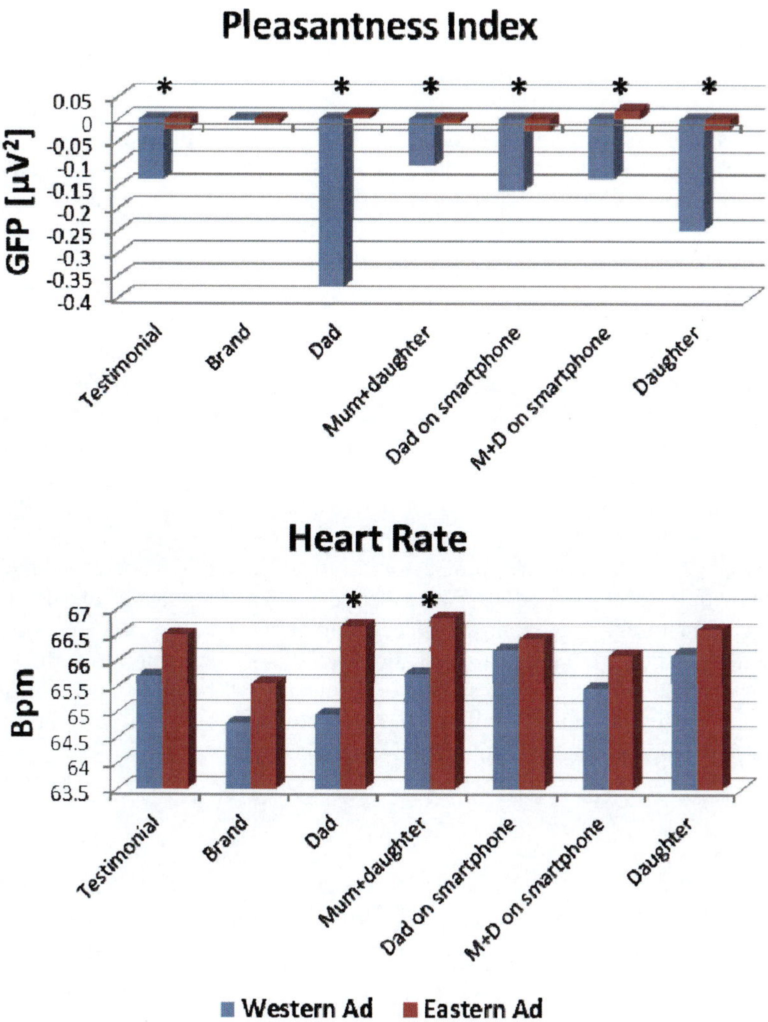

Fig. 32 Figure presents the average values of GFP for the Pleasantness Index (upper alpha, upper row) and Heart Rate values (lower row). Each column refers to a single segment of interest showing the values of the cerebral variables for both Western (blue) and Eastern (red) advertisement. The differences of the Pleasantness Index are statistically significant for each segment except the Brand one (as the symbol * indicates). The differences of the Heart Rate values are statistically significant for the Dad and Mum + Daughter segments (as the symbol * indicates), with $p<0.05$.

Instead, the Eastern version of the ad is characterized by values of PI close to the zero and also positive ones. Overall, by comparing the PI values of the two versions of the ad, we can observe that the Eastern TV commercial is perceived more pleasant with respect to the Western one. The difference of the PI are statistically significant in each segment of interest except for the Brand one, where on the screen is presented the same logo for both advertisements. The level of statistical significance was taken as $p<0.05$ after Bonferroni correction with respect to the number of test performed.

As far as concern the analysis of the Heart Rate, the two commercials present the same HR values in almost all segments except for the two ones in which the only father appears on the screen and the one in which both mother and daughter act. In both cases the HR values are higher for the Eastern ad. This result suggest that these two segments of the Eastern ad have been perceived with a more positive emotion with respect to the same scenes of the Western ad.

Results above presented highlight how a population of Eastern people differently react, from a cerebral point of view, to the observation of an Eastern TV commercial with respect to a Western one of the same item advertised. In particular, from a cognitive perspective, we found out no difference in the Memorization and Attention Index between different scenes of interest of the two commercials. However, significant differences appeared in the emotional variables. Both Pleasantness Index and Heart Rate showed how the Eastern ad is perceived more pleasant and with more positive emotion (Vecchiato et al., 2011) with respect to the Western one. Present findings suggest how Eastern population investigated is more attracted from actors and situations they perceive more familiar with respect to ones presented in the Western version of the TV commercial analyzed. This could be of help for marketers since it seems to be important to adapt the commercial campaign according to the country in which it has to be promoted. Further analysis and experiments will be performed in order to better investigate cultural difference and similarity between Eastern and Western population during the fruition of TV commercials.

Chapter 8
The Added Value for the Evaluation of Marketing Stimuli

The aim of this Chapter is to show how it is possible to extract cerebral indexes which are able to describe memorization, attention and pleasantness processes running during the observation of ads. In particular, we show the existing linear correlation between the memorization index and the explicit report of subjects enrolled in the experiment. In addition, we also illustrate that the increase of cerebral variables of interest occurs in short time intervals. This information could be used by marketers in order to insert in this specific time slots the message to promote to people (Vecchiato et al., 2012d).

8.1 Electrical Neuroimaging Can Improve the Quality of the Marketing Messages

The already described selected theta and beta frequency bands have been employed to calculate the Global Field Power (GFP; Lehmann and Skrandies, 1980), and to define the Memorization Index (MI) and the Attention Index (AI), respectively. Such indexes are linked to the episodic memorization and attention posed to the ad by the experimental population, and they have been previously validated by using a correlation analysis with the answers the subjects gave during the interview. In this way, from the GFP waveforms we calculated the correlation coefficient between the MI and the percentage of spontaneous recall, as reported by subjects during the interview, as well as the duration of peaks elicited on such waveforms.

As far as concern the analysis of the Memorization and the Attention Index, on average we obtained values of MI = 0.37 and AI = 0.36 across all subjects, while the mean value of spontaneous recall is 31%, across all TV commercials. The results of correlation's analysis between the MI and the percentage of spontaneous recall are illustrated in Figure 33. In the scatterplot, the mean percentage of spontaneous recall is presented for each advertisement showed to the population, against the related value of Memorization Index. The percentage of spontaneous recall is then linearly correlated with the value of MI (R^2=0.68, $p<0.01$). These values are distributed along the regression line, as showed in Figure 33.

G. Vecchiato et al.: *Neuroelectrical Brain Imaging Tools*, BIOSYSROB 3, pp. 85 – 88.
DOI: 10.1007/978-3-642-38064-8_8

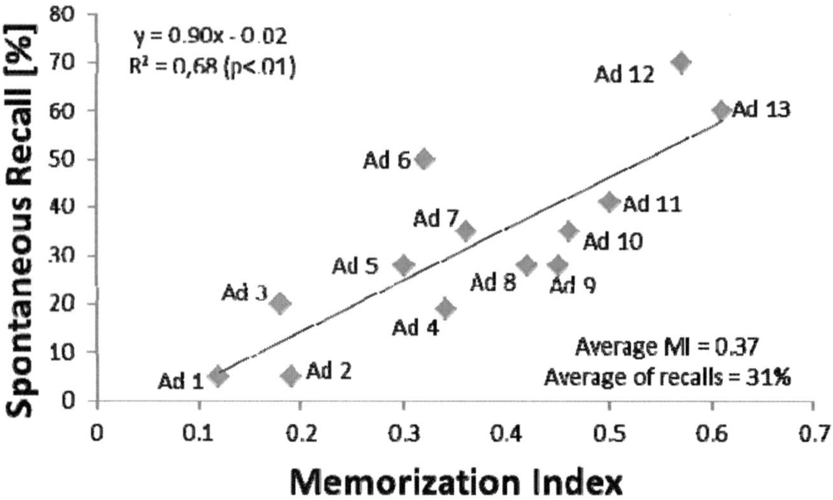

Fig. 33 In this picture the correlation results between the Memorization Index (MI, on the x-axis) and the percentage of spontaneous recall (on the y-axis) is presented. The Figure shows the linear correlation between the two variables along with summary of analysis' results and the average values of both MI and spontaneous recall.

Panel A of Figure 34 shows the percentage of spontaneous recall for different combinations of MI and AI. In particular, we report that when both MI and AI are under their average values the percentage of spontaneous recall (18%) is under average as well. This percentage is slightly increased (20%) when the AI exceeds the average's threshold. We obtained the highest values of spontaneous recall when the MI is over average. In fact, in this case the percentage reaches the value of 33% when the AI is under average and the value of 41% when both MI and AI are over average.

From the GFP waveforms we also calculated the duration of peaks. In panel B of Figure 34, we show that the peak's mean duration, across subjects and commercials, is of 2.3 s for those elicited in the beta band and of 1.2 s for peaks occurred during the observation of advertisements in the theta band, respectively associated to the activity of attention and memorization processes.

The analysis performed on the Memorization Index showed that it is possible to extract, from the EEG signals, cerebral indexes which convey information about the process of memorization of the TV commercial seen. The statistical analysis revealed a positive correlation between the MI and the percentage of spontaneous recall. The present finding confirms that an enhance of spectral EEG activity in the theta band is strictly connected with an increase of the probability to remember a commercial seen on TV. In addition, this probability also depends on the variation of activity of the beta band. In fact, we reported that an increase of the Attention Index contributes to enhance this probability. However, the present analysis also showed that only less than 1% of the GFP is involved in significant

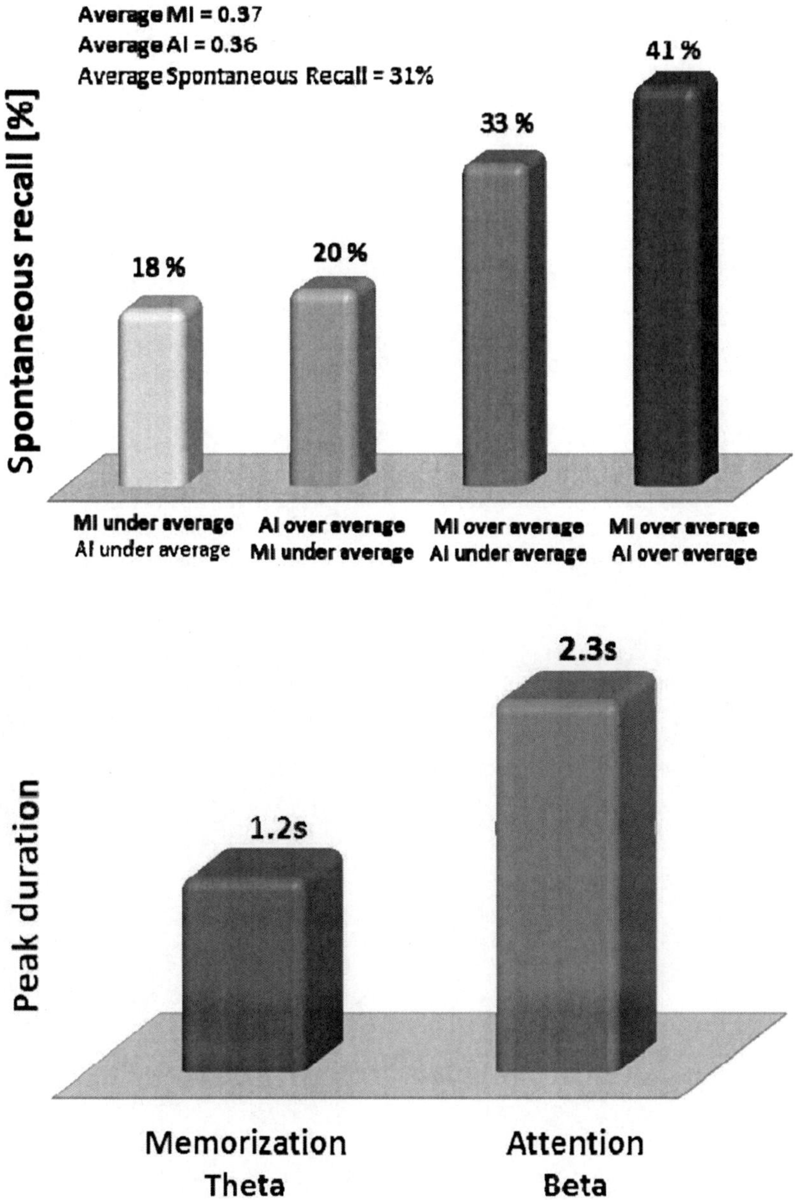

Fig. 34 Panel A shows the percentage of spontaneous recall when Memorization Index (MI) and Attention Index (AI) are above or below their average value. Panel B illustrates the average duration of peaks elicited in the GFP waveforms in the theta and beta bands, related to the memorization and attention process respectively.

peaks of activations. This information could suggest marketers where to insert key frames within the commercial message since the memorization and the attention processing fill just very short time intervals of the advertisement.

Taken together, these mentioned examples also bring evidences that it is possible to link some properties of the collected EEG rhythms during the watching of some TV advertisings with the overt preferences of the observers in terms of emotions.

We showed that this tool is able to help us in observing and analyzing the temporal evolution of memorization processes thanks to the high temporal resolution. This feature allow us to distinguish frames and scenes of interest by means of a graphical representation of the temporal waveform of the average GFP. In such a way for each film segment of a clip it is possible to distinguish peaks of activations elicited.

In the present Chapter we illustrated how, by means of appropriate mathematical and statistical methods, it is possible to gather information hidden in the cerebral activity about the memorization of a TV commercial. It is worth of note that improving the quality of the marketing messages allow the industries to lose less money in the production of inefficacy or incorrect advertisements and finally help them to match better the demands of the people related to the good to be advertised.

John Wannemaker, the inventor of the first mall in US, said "I know that half of my money spent in advertising is wasted, but I don't know which is such half". The use of neuroimaging tools in the evaluation of the commercial ads could help to reduce the half part of the money that were wasted in advertisement industry, by preserving the correct one.

Chapter 9
Application to Real Cases

9.1 Neuroelectrical Imaging Can Improve the Quality of the Marketing Message: A Case Study for the TV Commercial "Impresa Semplice" by Telecom Italia Broadcasted in Italy on March 2012

In this chapter, an example of the practical application of the previously discussed neuroelectrical imaging indexes is provided for the evaluation of a commercial advertisement before its TV launch on air. The aim is to describe which kind of additional information the analysis of cerebral signal could bring to the industry in this area of research.

The discussion of the following example could be made due to the graceful written permission of Telecom Italia, who is the main telecommunication company in Italy that commissioned the presented study. Telecom Italia is leader in the TLC field and, periodically, proposes various offers for fixed and mobile telephony and innovative information technology solutions. The analyzed commercial, described in this chapter, could be found (on its final broadcasted form) on YouTube at the following http address: http://www.youtube.com/watch?v=AfEkECgbvsI and it is part of a long communication campaign promoted by Telecom Italia known as "Impresa Semplice" (Easy Enterprise). In particular, Telecom Italia commissioned a neuroelectrical imaging pre-test in Italy on March 2012 on two similar versions of the same advertisement for "Impresa Semplice", one of the Telecom Italia brands. Such a service proposal was destined to the telecommunication field of small and medium-sized enterprises.

The two main goals of the test were: 1) to analyze the TV commercial efficacy frame by frame and 2) to find out, if any, eventual differences in the perception of the brand showed in the last seconds in two different ways. Therefore, the commercial was tested in two versions differing only on the short brand frame sequence at the ending. These two versions were slightly different from the YouTube version of the spot that was broadcasted after having implemented some of the suggestions highlighted by the neurometric analysis that it is going to describe in the following.

G. Vecchiato et al.: *Neuroelectrical Brain Imaging Tools*, BIOSYSROB 3, pp. 89–112.
DOI: 10.1007/978-3-642-38064-8_9

The "Impresa Semplice" campaigns were characterized by the presence of a stable testimonial, a simple layman that was previously unaware to the general public, thus not being a "movie star". The general idea was to force the identification between the normal layman and such testimonial. A characteristic trait of these series of commercial advertisements for such campaign was the contraposition between people using the Telecom Italia "Impresa Semplice" phone services, identified by a "red right arm", and the testimonial, identified by the layman without the right arm of the coat, who doesn't use the services by "Impresa Semplice". The key theme in the analyzed commercial is **the proposal of a solution to the problem of the high** costs for the call from a fixed to a mobile telephone. The analyzed ad starts with the testimonial that gets out of his office bringing his telephone to avoid that anyone makes expansive calls to mobile phones while he is far away. Once he leaves his not modern equipped office, he discovers a different world: the one of professionals choosing "Impresa Semplice". *"Il braccio destro che fa per me" (i.e.: "The right arm, right for me")* is the advertising claim that sounds so simple and efficient joining the name of the proposed new offer "Linea Valore +" (i.e.: "Product line Value +"), subscribing that the call costs from fixed to mobile telephone will be no more a problem.

Before the description of the results starts, it is important to highlight the analyzed TV commercial creative schema. The ad has first an introductive phase (1''-8''), thus there is the longest part (from 9'' to 24'') in which the specific offer of the services provided by Telecom Italia under the Impresa Semplice name is presented, then the presentation of the brand (25''-26''), and finally a brief return to the story in the last part of the clip (27''-30''). Such configuration is quite original because most of the commercials normally broadcasted in Italy have a simpler structure, with a sequence of three only phases: a story-telling, the proposition of the offer and, successively, of the brand.

Figure 35 shows the temporal segmentation of the analyzed advertisement which has been considered for highlighting the impact of the different phases of the frame segments, in the clip perception analysis.

In order to analyze the impact of the ad, on the base of the cerebral activity, the video was logically divided in several phases, named frame segments. In Figure 35, the temporal duration of each one of these segments is represented by a colored line below each frame of the broadcasted video. The temporal segments have been chosen as follows: *Introduction,* where the testimonial appears (the layman without the "red right arm") while is getting out from his office carrying his telephone (1''5''); the *Central story* section, characterized by the presentation of the service "Linea Valore +" (from 9'' to 16''); the *Brand* section, with the appearance of the Telecom Italia brand which also represents the only difference between the two versions in test (25''-26''); the *"Tail"* segment closes the announcement recalling the narrative theme (27''-30'').

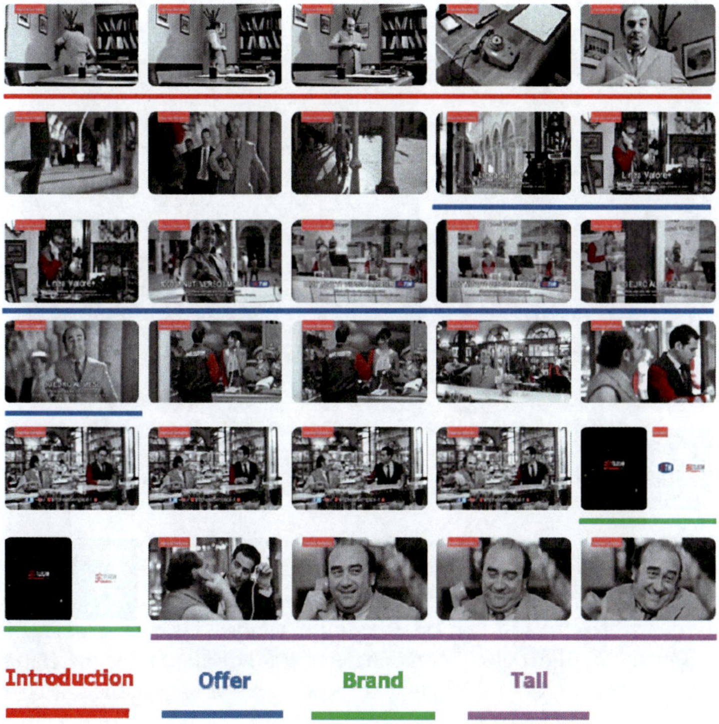

Fig. 35 Frame and temporal segmentation of the TV commercial for main interest segments.

The recording of the biometric response included the detection of the EEG signals, HR (Heart Rate) and GSR (Galvanic Skin Response) parameters on a sample of 24 subjects (12 men), aged between 35 and 54 years, in two Italian cities: Milan and Rome. Subjects belonging to the target group were owners or partners of small entrepreneurial and professional activities (from 1 to 9 employees). Moreover, the efficacy and the impact of the spot were measured on the total collected sample or between the different subgroups, derived from it, defined as follows:

1. *Users*, already Impresa Semplice's customers, TLC Business Telecom;
2. *No Users*, customers of other mobile operators.

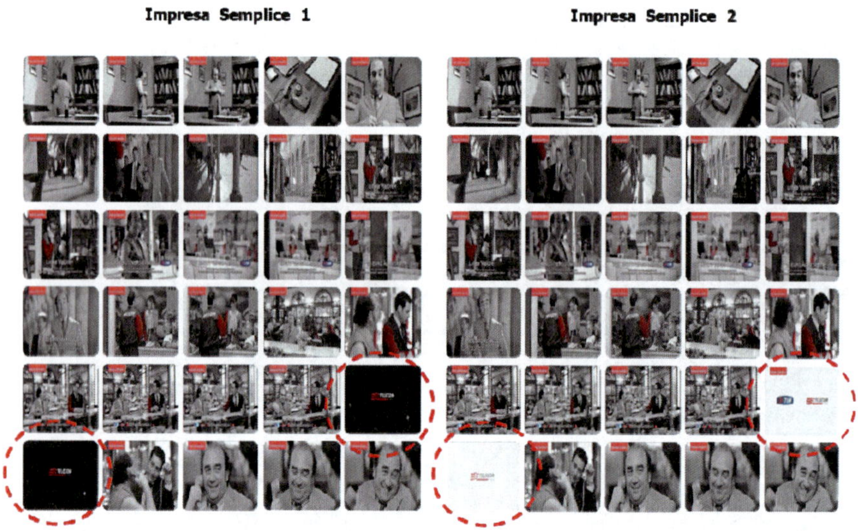

Fig. 36 Frame sequence of the two analyzed Telecom Italia TV commercials. The two clips, named IS_1 e IS_2 visually differed just for the background of the proposed brand.

As previously described, the test was relative to the analysis of two versions of the same ad, differing in the color of the background of the Telecom Italia proposed brand (e.g. black or white). Figure 36 shows the frame-sequence of two versions in test, differing only in the brand presentation at 25''-26''sec.:

1. In the first version (IS_1), Telecom brand appears on a black background for 2 seconds;
2. In the second version (IS_2), Telecom brand appears on a white background in two different ways: one second with three logos "**Impresa Semplice** + **Tim** + **Telecom**", the following second with only the **Telecom** logo (Figure 36).

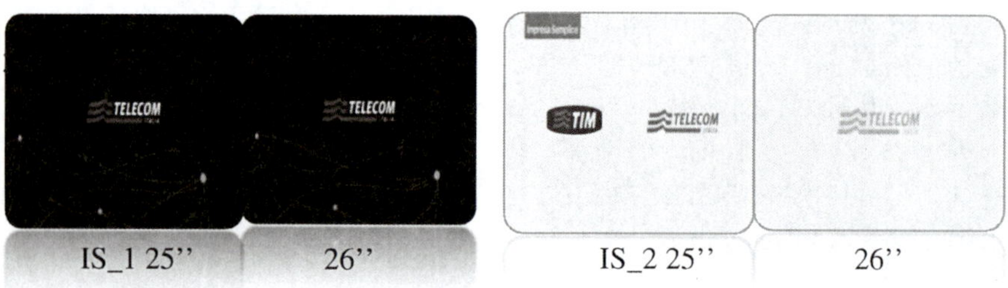

Fig. 37 In this Figure the differences between the "Brand" proposition between the two versions of the analyzed spot (IS_1 and IS_2) are displayed.

The analyzed advertisements were immersed in a commercial break inserted during the observation of a documentary. The entire analyzed sample viewed the documentary and successively the proposed commercial break with different commercial clips within, including the analyzed ones.

After evaluating the biometric response of the analyzed subjects for the two ad versions, it was possible to assess the impact of each one, the efficacy of the proposed offer and the impact of the brand in the two different versions.

9.2 A Neurometric Analysis of the Impact of Creativity and Measurement of the Efficacy for Particular Time Segments of the Analyzed Advertisement

The biometric EEG, HR and GSR recordings returned neurometric indexes linked to the variation of the memorization, attention and emotional involvement of each analyzed person. Variation of such indexes along the adopted time segments returned information about the ad perception, second by second, on the total sample of recorded subjects and on sub-samples of such group. In particular, the total analyzed group could be further divided in subgroups. They are composed by different gender (Male/Female) or by particular characteristics relevant for the analysis (such as User / No User for the particular good or service offered in the advertisement, or other demographic aspects).

The neurometric indexes related to the fluctuations of the memorization, attention and emotive processes of the total sample along the ad perception could be also composed in a unique indicator, named "Key Frame", conveying the grouped information from all the other indexes. This was made to facilitate the interpretation of the cerebral reaction of the analyzed population to the TV commercial perception. In fact, the Key Frame index is obtained by combining memorization and emotion signals.

In Figure 38 the average time-varying values of the 3 indexes related to memorization, attention, emotion and the derived key frame, for the two ads in test are plotted. The indexes are normalized with respect to the documentary level. **Thus, the indexes are expressed in z-score values.**

By observing the different plots, the congruence between the value of the indexes for the two analyzed spots (IS_1 and IS_2) appears clear. This was substantially expected, being the two versions of the spot equal for almost the whole film, except for the tail where the two spots were different. As we can see in the Figure 38, the ad perception generated variation of the indexes, in the total analyzed sample, that are positive (e.g. higher than those observed during the documentary) for almost the entire clip duration.

It is possible to see that the attention index trace is not particularly high but, memorization and emotive involvement levels are definitely above the relaxing/neutrality condition produced by the documentary.

It could be appreciated, by the observation of Figure 38, as all the indexes obtained for the whole analyzed sample are lower in the first 10'' of the commercial when compared to the values in all the other part of the it. This seems to suggest as the initial part of the two ads were not particularly appreciated by the investigated population.

The presence of the man without sleeve, without dialogue and with the only communication by music and imaging, initially captures the attention of people/audience, but soon after, during the walk under the arcades, there is a decline also more evident in attention. In the same seconds also emotion is low.

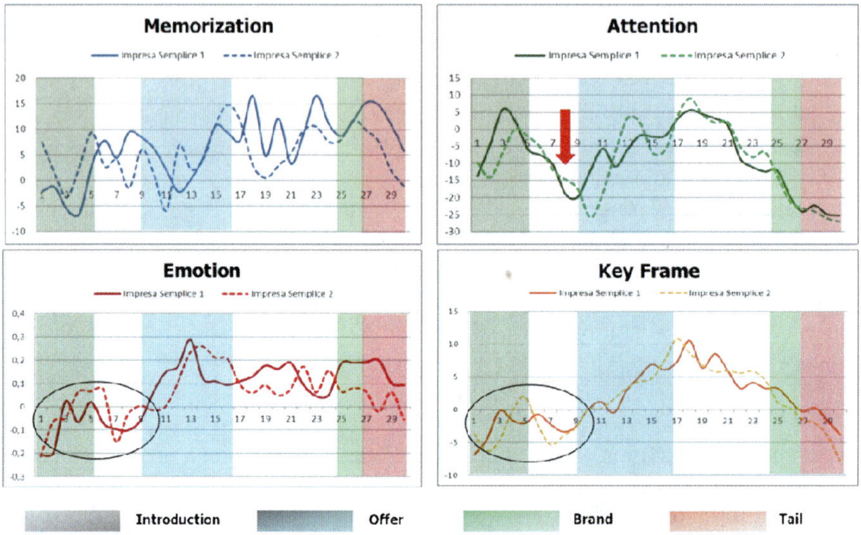

Fig. 38 Cognitive and emotional trends of the analysis sample, during the observation of the two investigated TV commercials (Impresa Semplice 1 and Impresa Semplice 2). Shaded areas show the time period of particular segments of the spot such as "Introduction", "Offer", "Brand" and "Tail". The continuous lines are referred to the computed indexes for the whole sample for the Impresa Semplice 1 ad, while the dashed lines are relative to the computed indexes for the Impresa Semplice 2 ad. The horizontal axis describes the time evolution of the clip (from 1 to 30 seconds) while the vertical axis describes the z-score value of the considered index for Memorization, Attention and Key Frame. Vertical axes for Emotion indicates positive emotion from 0 to 0.5 values and negative emotions for values from 0 to 0.3. Such last values are not expressed as z-score. Circles drawn on the Emotion and Key Frame lines highlight the low values of such indexes at the beginning (first 10'') of the video-clip in the whole sample.

The neurometric indexes increase during the central part of the considered advertisements, in particular during the period in which the publicized service was displayed ("Offer" segment). Such part is depicted in shaded color in the Figure 39. It is characterized by a late increase of memorization index, which improves in the last part of such segment. This last part of the segment of the advertisement is characterized by a voice that describes the economic advantage of choosing that particular offer for making phone call (1000 minutes of phone call for 30 euros per months). It is the part of the spot where the economic arguments hold.

Precisely, during this presentation of the "Offer", there are 3 changes of written statements: the first shows a short phrase "Linea Valore +" (name of the offered services) and is on for 3 seconds; the second shows "1000 min versus TIM mobile" for other 3 seconds; finally, the last writing in overlay is "30 euro to months", plotted for 2 seconds; in addition, under every sentence, there are always normative rules, written in small letters, and contractual terms without the speaker always pronounces as mentioned in written.

During the same time period, the attention index has a behavior similar to the memorization one, while the emotion index runs in positive emotional area, reaching its maximum across the entire spot.

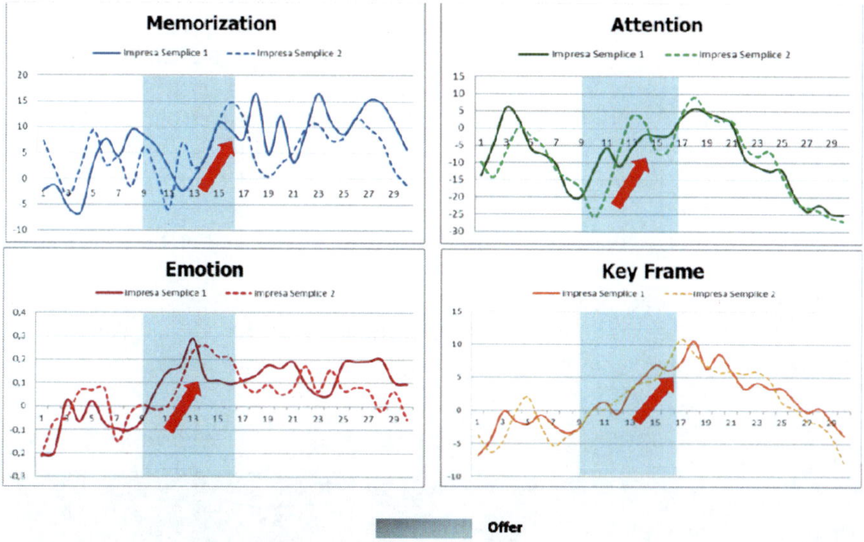

Fig. 39 Cognitive and emotional trends of the whole sample relatively to the "Offer" segment of the analyzed TV commercials. Same conventions of Figure 38.

Such positive emotion remains substantially elevated (between 0.1 and 0.2) until the end of both clips (as described by the continuous and the dashed lines in Figure 39).

Summarizing, the segment of the advertisement, in which the specific offer is presented, generated good values for emotion, attention and memorization, especially in the last part of it. This last part is where the most economical arguments were presented.

In the performed neurometric test the information related to the cognitive and emotive perception of the clip, were collected not only by biometrics analysis but also by using a semi-structured questionnaire, administered by an interviewer at the end of the recording session to the persons who participated to the data collection.

The collected verbal declarations show coherent information when compared to the development of the computed cognitive signals during the offer time segment of the ad. They are characterized by a low memorization for the first seconds and a higher memorization at the end.

In particular, the analysis of the questionnaires revealed that all the interviewed persons are aware of having watched such TV commercials. However, nobody of them remembered spontaneously the name of the service presented in that period of time in the clip ("Linea Valore +").

The name of offered service ("Linea Valore +") appeared written on the screen only for 3 seconds, during the "Offer" segment, without being ever pronounced by the speaker. There are experimental evidences that suggested as the written text could be better understood during the clip if the speaker pronounces it. Thus, it was not surprising that no people remembered the name of the particular promoted offer in that part of the ad. However, many people of the analyzed group remembered other particulars of the TV commercial, for example the economic information about the offer (e.g. 1000 minutes of phone conversation at 30€ per months). In fact, such information were presented as a written text in the clip, but also simultaneously pronounced by the speaker.

It also happens that verbal answers, given in the brief interviews following the biometric recordings, consisted in verbatim that were focalized on the informative and commercial of the clip. These aspects were often confused and sometimes irrelevant with the real content of message.

This happened because the several information provided by the speaker and for the presence in sequence of too much written signs in overlay and also not always corresponding to the speakers' words at the same time.

Fig. 40 Segmentation of the presented "Offer" for the TV commercial analyzed.

Since the experiment was conducted in pre-test, from the detailed analysis of the biometric perception it has been possible to indicate potential improvement interventions both for creativity and for the offer, specifically.

Considering the difficulty of memorization and attention, especially during the first part of the advertisement, it seemed necessary to adopt measures to contain the attention decay after the first seconds of the clip. Such measures were adopted in order to simplify the decoding of the message during the offer time segment and to put out, in different and easier way, the crucial informative elements to facilitate the people understanding.

Practical suggestions for this first part of the ad, on the base of the analysis of neurometric indexes, have been:

- A possible reduction in the narrative part of the first 10 seconds to avoid a marked decrease of attention before the offer;
- The addition of a narrative voice during the first part of the offer in order to communicate more clearly the written name of the service "Linea Valore +";
- A simplification of the information, contained in the offer time segment, to aid the understanding of some details of it.

9.3 Analysis of the Impact of the Brand and of the Narrative Tail in the Two Different Versions of the Spot Tested

The two analyzed video-clips differ for the part relative to the brand proposition, with black and white background (Figure 36). These differences in stimuli produced, in the analyzed subjects, the differences in neurometric indexes highlighted in the Figure 41.

In particular, it was noted a significant difference for the emotion index, which is more positive for the version 1 of "Impresa Semplice" (S_1 "life network" black background). All the analyzed cognitive indexes, instead, resulted similar for both commercials (with only a small difference in memorization in favor of IS_1, black background).

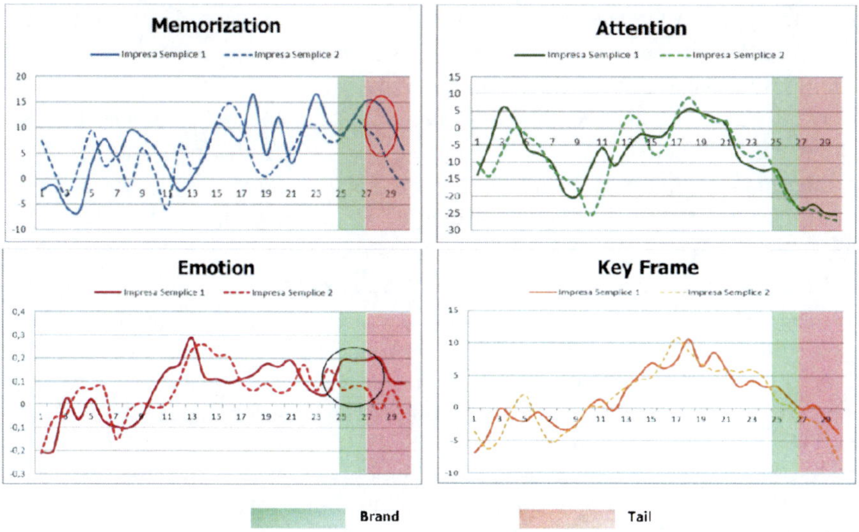

Fig. 41 Cognitive and emotional trends of the sample specifically in "Brand" and "Tail" segments, represented by the shaded areas. Same conventions as in Figure 38.

After the occurrence of the "Brand" segments, in the short tail of the storytelling following the brand (pink colored part in the Figure 41), both cognitive and emotional indexes decreased for both ads. However, the memorization and emotional indexes are slightly better in the version Impresa Semplice 1 (in continuity in the different effect on the brand) than for the second version of the it, Impresa Semplice 2.

A reasonable interpretation of such neurometric finding is that the appearance of the brand suggests to the analyzed population the "end" of the spot. The successive restart of the narration (the "Tail" segment) is not able to restore appreciable levels of the neurometric variables such as attention, memorization etc.

Thus, it was suggested to remove completely the tail section of both advertisements in order to gain seconds to further elucidate the offer section of them.

9.4 Analysis of the Perception of the TV Commercial "Impresa Semplice" in Two Sub-groups User and No User of Telecom Services

In the contest of the study commissioned by Telecom on "Impresa Semplice" commercial, an analysis of difference in the perception of the videos, between the two sub-groups of Users and No Users of Telecom services, was also performed. These two sub-groups often correspond to different areas of marketing strategies, and then their different perception is particularly interesting. In this case, the analysis brought out that the two groups were different in their biometric responses for the attention and emotion indexes.

With respect to the memorization index it is quite similar across the two subgroups for both the two clips versions, except in the final part. In fact, in the version Impresa Semplice 1 (indicated in Figure as IS_1) the memorization index is higher for the Users group on the final "Tail" segment than for the No Users one.

As far as the attention index concerns, results suggest as in both versions the No User subjects are more attentive than the User subgroup (customers Telecom), especially during the final part of the advertisements.

Also for the emotional index, a significant difference is visible between User and No User subjects during all the commercial observations. In both the ad versions (indicated in Figure as IS_1 and IS_2) the Users subgroup has been more involved (i.e. with higher values of the emotional index) than the No Users subgroup. These differences are visible along almost the whole spot and in particular during the second part of it, particularly at the appearance of the Telecom brand and during the following closure ("Tail"). **This effect is in agreement with the idea of a positive emotional recognition by the Users subgroup of their own brand/company.**

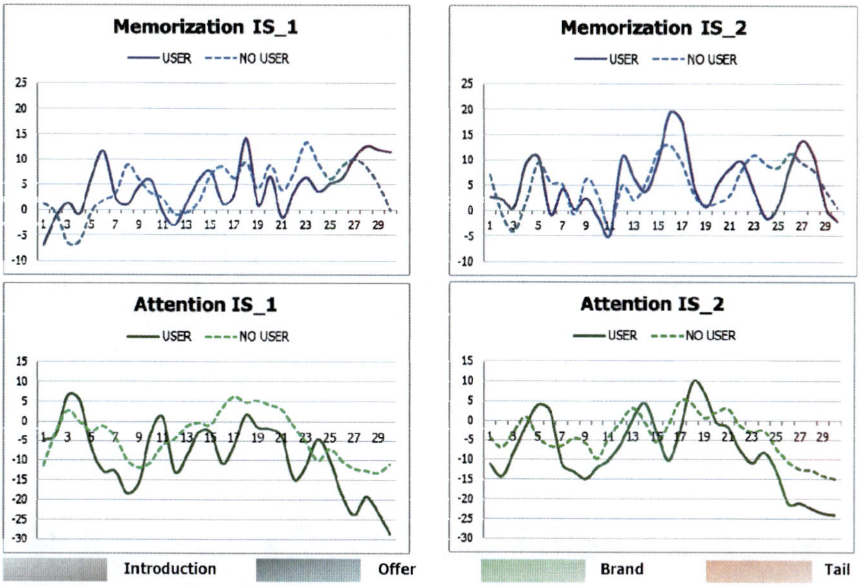

Fig. 42 Variation of the Memorization and Attention indexes along the two analyzed video-clips for both the investigated sub targets (User – No User). Same convention of the previous Figures. The dashed line is relative to the No User subgroup.

Regarding the difference in the subjects' response of the two different "Brand" segments, although the difference is minimal, Telecom Users responded emotionally better at the white background "Brand" segments (see IS_2 in Figure 43, sec. 25''-26''). Otherwise No Users were clearly much more involved by the black background "Brand" segment (see IS_1 in Figure 43, sec. 25''-26''). On this black background "Brand" segment, they had positive emotional values versus negative values induced by the white background one.

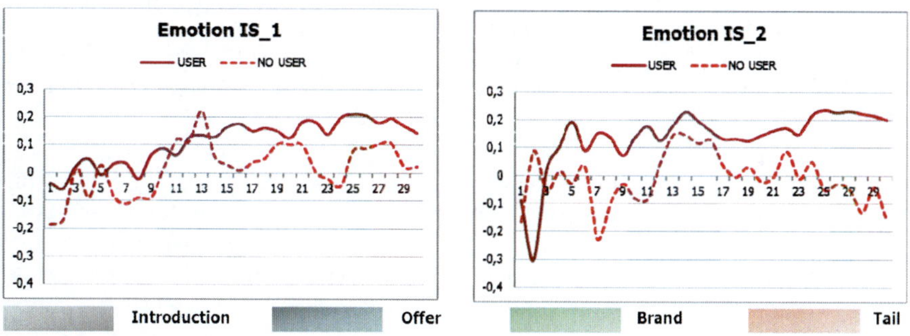

Fig. 43 Emotional trends for the sub-target User - No User. Same convention of the above Figures.

9.5 Implementation of the Neurometric Test Results for the Final on Air TV Commercial Version

The findings obtained by the neurometric pre-test have been taken into account in the production of the final version of the TV commercial "Impresa Semplice" really aired by Telecom Italia. The main modifications to the pre-tested versions have been made in the "Offer" segment and in the "Tail" one.

Firstly, the voice of the speaker and the presentation of the "Offer" have been postponed forward, from 9'' at 13'', after the scenes that during the pre-test analysis resulted with high level of attention index. Moreover, the scenes and the speaker voice have been better coordinated with the appearance of the written statements on the screen, in order to maximize the synchronization of the visual and auditory perception.

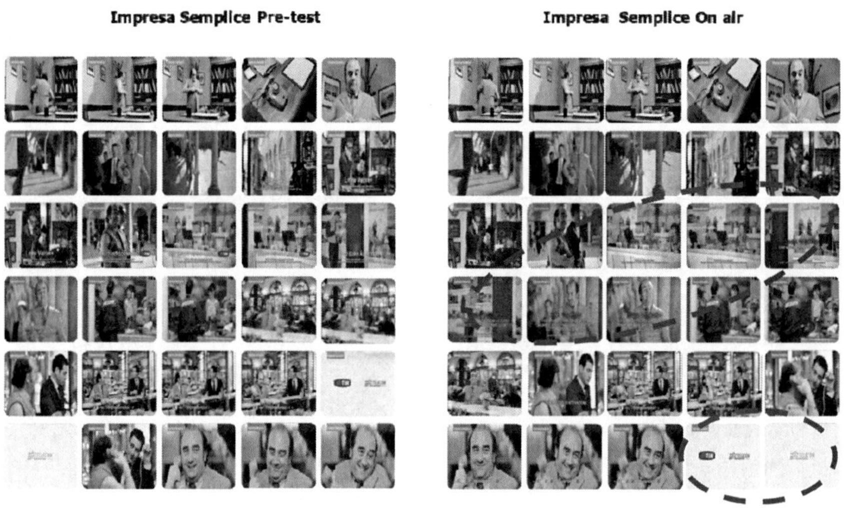

Fig. 44 Comparison between pre- test and on air advertisement. On the left, the sequence of a particular video-clip tested with the neurometric analysis (white brand). On the right, the sequence of the advertisement effectively launched on air. Dashed lines highlight the performed changes of the video frames composition in order to make more clear the message of the spot.

In the frame sequence of the "Offer" section, it has been clearly highlighted the name of the service "Linea Valore +" both for visual level (with words presented in red color) and for audio level (introducing the speaker voice where in pre-test video was missed). The service name "Linea Valore +" has also been kept in overlay for the whole presentation of the "Offer" segment. The information "1000 min" and of "30€ per months" are successively added on the screen and hold as written text under the name of the service "Linea Valore +", while the speaker's voice lists them clearly. In this way, in the final aired commercial each detail became clearly easier to be perceived by the subjects.

Fig. 45 Offer sequence by "Impresa Semplice" in the final on air release.

In figure 45, is visible how, for the final release of the clip, the name of the service (highlighted in red) remained in overlay on the screen for all the 6 seconds of presentation of the "Offer" segment. Moreover, each additional information (1000 min versus TIM, "Fixed calling national unlimited", "30€ per months") were also inserted under the name of the service, remaining unchanged across all the "Offer" segment duration.

In addition to the changes operated on the "Offer" segment, the neurometric insights offered the possibility to change the appearance of the "Tail" segment. In fact, results from neurometric pre-test highlighted that after the presentation of the brand in the ad, all the cognitive indexes related to attention and memory suffered a continuous decline. Thus, it was decided to anticipate a shorter version of the previous tail by continuing the former narration of the spot, and the aired version of the spot finishes with the appearance of the brand.

The two versions of the TV commercial pre-tested differed in the background color of the "Brand" segment: what's about the final version of such color chosen for the aired version?

In the final aired version the company chose the white background brand frames like in version Impresa Semplice 2. The two versions were not sharply different in their cognitive response (see Figure 41) but the chosen white one generated higher and maintained values in memorization index for the Non User subgroup (see Figure 42). This group was the main target of the proposal, and also the difference between User and Non User was statistically significant.

9.6 Automotive Sector, Case Study of Alfa Romeo Giulietta Commercial with Different Impact in Emotion for Men and Women

The difference between men and women is an established fact. Also in neurometric test, the perception of the same communication may be also sensitive to differences between male and female. The following example shows a neurometric test on a TV commercial for a famous Alfa Romeo car model, performed for scientific proposal at the EEG Labs of the "Sapienza" University in Rome. Hence, such test was not commissioned by the owner of the Alfa Romeo brand. The analyzed TV commercial could be found at the following address on YouTube:

http://www.youtube.com/watch?v=ydhRH_LUAk4

The TV ad aimed to convey a strong emotional message related to the new car model "Giulietta". Such car is proposed not just “as an Alfa Romeo car” but as a car suggesting sensations of elegance and security that would like to be transmitted also to all the future car buyers.

The protagonists of this new car advertisement are the beautiful Uma Thurman (at the driver’s seat) and three blond children in the back seat; a William Shakespeare’s quotation is the final claim: "*We are such stuff as dreams are made on*", which emphasizes even more the spirit and the atmosphere that this commercial wants communicate in this advertising campaign. The creative idea is the combination of the Giulietta car style, promoted as determined, affordable and high performing, with the personality traits of Uma Thurman, suggesting both determination but also beauty and child safety (as implicitly evoked by the one of the most famous role of Mrs. Thurman in the “Kill Bill” movies series).

The scientific study investigated the cerebral activity of a sample of subjects while they were watching a documentary intermingled with a series of advertisements including this one. Electroencephalographic (EEG), GSR and HR recordings were carried out on a sample consisting of half men and half women. In Figure 46 the segmentation frame by frame of the commercial is shown.

Fig. 46 Frames sequence of the Alfa Romeo Giulietta TV commercial.

Results of the Physiological Responses. From the analysis of the indicators of memorization, attention and emotion activities (shown in Figure 47), it was evident that the commercial video captured the attention of the involved subjects. In particular, there are frames of the commercial that were high in those cerebral indexes that helped to strengthen the image of **serenity** (1" -2") and **peace** (7") conveyed by the new advertised car. Moreover, the final frames have high value of the cerebral indexes reflecting interest and curiosity to try the car by the investigated sample (20" -30").

The advertisement presents the main peaks of the memorization index during the scenes underlining the main contents of the message: the **pleasure of driving** (1"), the **safety** of the car (10" - 11" with the message -The safest in its own category-), **the name** and the **elegance** (16") and the **payoff** (30"). Moreover, it resulted as a clip dragging the audience on a higher emotional level especially during the first scenes, when the car is driven and when the three children are shown. The cerebral indexes showed no substantial responses during the proposition of the W. Shakespeare’s citation (sec 16’’-20’’), and the same it is true for the relatively low emotional score observed. The final time segment of the ad (sec. 21’’-25’’), characterized by fading text describing the car engine qualities, results in a moderate peak of attention, immediately followed by a decline of memorization and emotion (Figure 47). One possible interpretation of this behavior of the indexes is the difficulty to decode the proposed message.

From these last considerations, it appears that in the investigated sample the final of the Giulietta advertisement is not particularly convincing and effective.

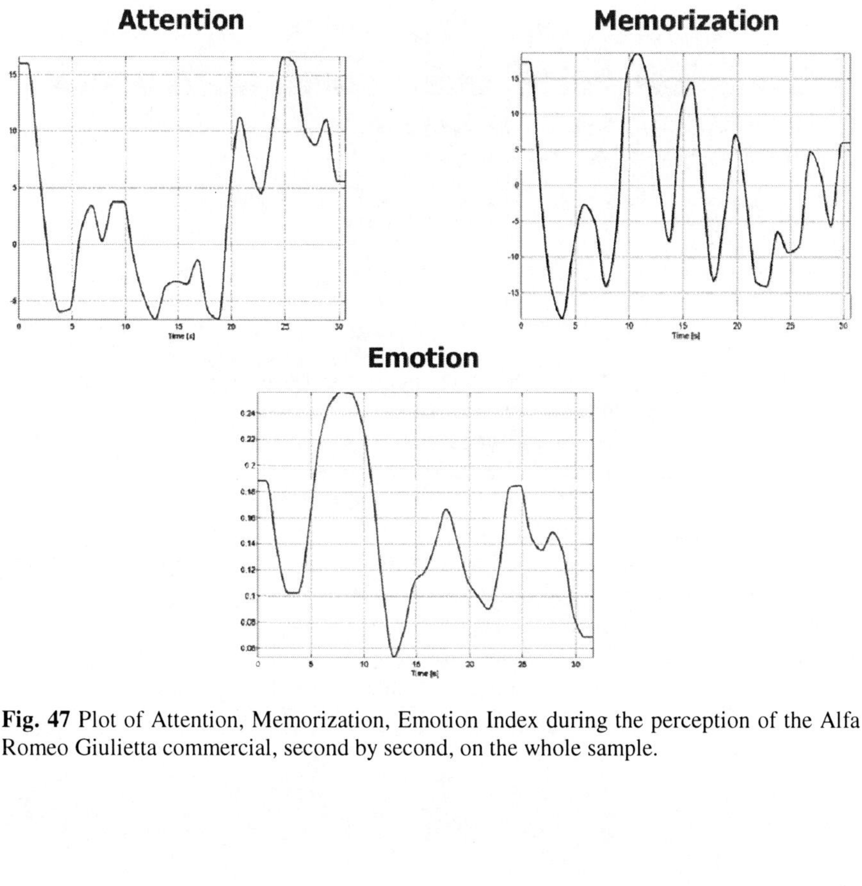

Fig. 47 Plot of Attention, Memorization, Emotion Index during the perception of the Alfa Romeo Giulietta commercial, second by second, on the whole sample.

By taking the frames in which the best or the worst performances of the computed indexes are shown, it is possible to overall determine the best and the worst concepts conveyed by the clip for the analyzed sample. In particular, Figure 48 suggests as the key frames for this ad are those related to the movement along the bridge, the children view inside the car in the backseats, the Alfa Romeo brand. The worst frames (from the point of view of the cerebral indexes) are those related to the Shakespeare's citation and the information at the end of the spot. Such frames of success and weakness of the TV commercial, on the basis of biometric records, were listed in the following Figure.

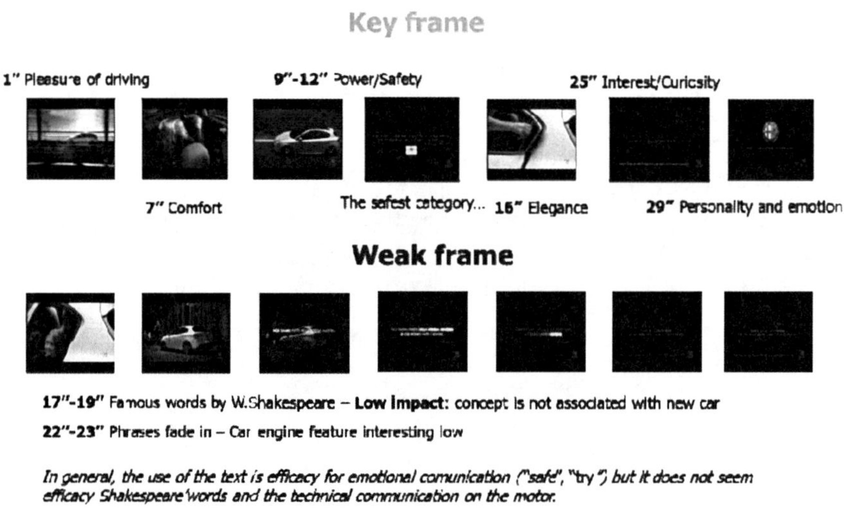

Fig. 48 Key Frame and Weak Frames of the Alfa Romeo Giulietta advertisement.

Further considerations could be made if we would like to pay more attention, in this example, about the neurometric application to the commercial ad analysis to the emotional index. This index has positive values throughout the whole clip, with several positive peaks in the range of values corresponding to the emotional states of "calm" and "serenity" for the whole considered sample.

Figure 49 shows the most positive peak of the emotional index occurring between the 6" and 10". In particular, on the scenes the spot testimonials appeared (Mrs. Thurman and the three little girls). After, on the successive scene there are other two additional emotional peaks: one when the Giulietta car ("the product") appears (sec. 17''-18''), and the second one at the beginning of the final claim (sec. 23''-24''). The TV commercial, altogether, has generated a positive emotional response. However, higher emotional values have been observed in the first part rather than in the second part of it, which is devoted to the presentation of the product.

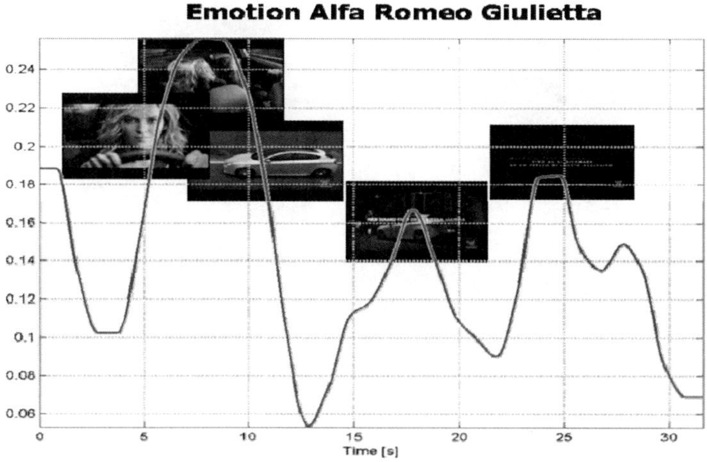

Fig. 49 Emotional profile of the Alfa Romeo Giulietta advertisement, sec. by sec.

However, this interpretation of the emotional profile of the whole investigated sample is going to change if a gender analysis will be made, as represented in the Figure 50. It was rather evident as the emotional profile derived from the Giulietta video is gender sensitive.

In fact, in the Figure 50 two distinct profiles for the computed emotional index are visible: one related to males and the other to females; they have just opposite trends. In particular, men were constantly captivated throughout the entire clip, with a greater emotional intensity during the final part of it. Instead, women have been highly involved during the first half of the ad, while their emotional response decay is almost linearly in the last ten seconds. Interestingly, the women's emotional index is high during the part related to the vision of the three little girls (6" -9") and on the images related to safety (10" - 11").

Thus, it appears, from the analysis of the computed emotional indexes, that the emotion for the men seems to be mainly conveyed by the final written message enhancing the car high technical performance, as well as for the Alfa Romeo brand. Instead, for the women, the emotion is captured by the female Figure and by little girls (while the speaker suggests the high safety of the car) as synonymous of beauty, tranquility and security. For the women it appears that the successive rational components of the message regarding the performance of the car engine are the less emotionally considered when compared to the male sample.

In such situation, it would be then possible generate short version of the previous commercial ad, differentiated by gender, to be aired in particular occasions as suggested in the following paragraph.

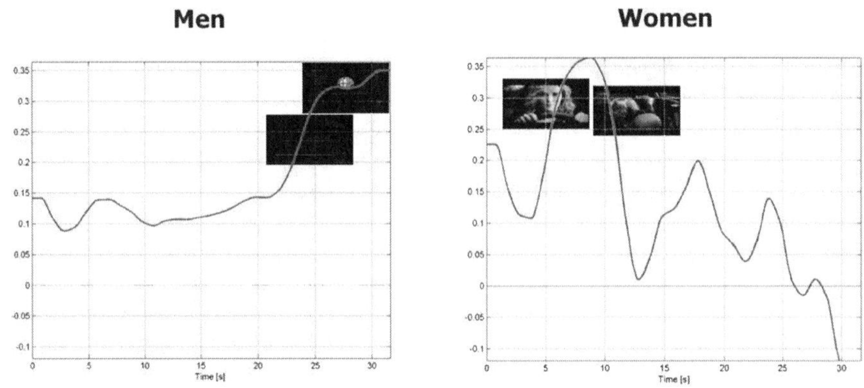

Fig. 50 Emotional profile of the Alfa Romeo Giulietta TV commercial in the sub-targets men-women. Left: emotional profile for the man's sample. Right: emotional profile for the woman's sample.

9.7 The Use of Biometric Techniques for a Possible Reduction of the TV Commercial Time Length, Taking into Account the Differences between the Male and Female Groups

Neurometric recordings provide signals of Memorization, Attention, and Emotion from which it is possible to evaluate the perception of the advertisement, second by second, as a whole and with reference to their individual segments that compose the creativity. This neurometric analysis provides also a rational schema useful to guide a reduction of the ad in time, such it is often necessary after the first creative production. The previous commented Alfa Romeo Giulietta commercial provides an interesting example of this rational reduction which can be applied to the video-clip as a whole, or even to produce different reduction more suitable to different media programs, oriented to different main targets.

By the analysis of the neurometric indexes plots, the frames on second most performing may be identified in those capturing more attention, more memorized, and the ones involving the viewers with higher emotional levels. These frames define a sort of path for an efficient cutting from a longer spot, by producing a shorter and more efficacy version of it. In this example for the Alfa Romeo Giulietta, the observed best key-frames were the ones showing the strengths of the car (name, safety, stability, power, comfort, elegance and sportiness). Even 15 seconds may be enough for composing such key frames, in a shorter clip than the previous one, but nevertheless still good to communicate the same features of the car to the chosen target. Maximum effectiveness is ensured by making it on the basis of the neurometric evidences. The following Figure just shows a reduction in frames, built from the evidence obtained by the neurometric test on the whole target.

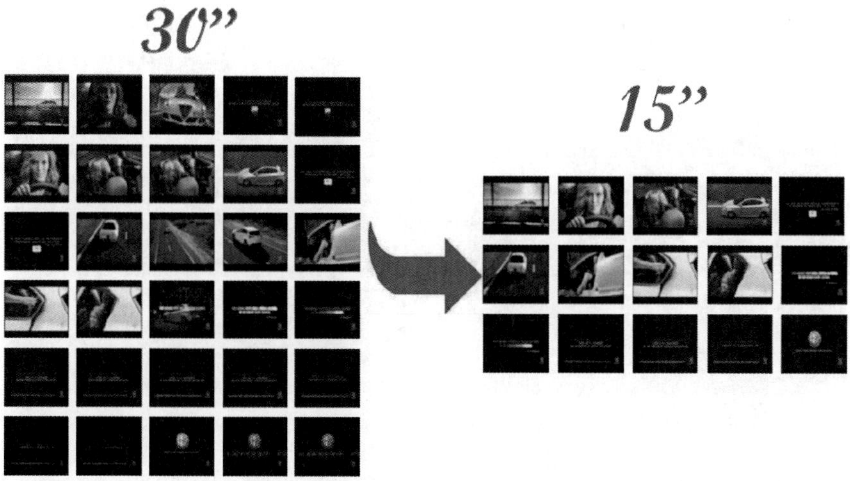

Fig. 51 Suggestions for a possible time reduction of the TV commercial to be applied to the whole analyzed sample (i.e. including males and females).

Nowadays it would be possible to broadcast a reduced version of the spot more focused on male or female subjects in different media contests. Considering the different perception of the commercial between men and women, it has been performed also a shortcut "focalizing and personalizing" the ad according to the male or female targets. Then, the company would be able to decide which of the personalized advertisements better fits with each scheduled media program, thus, making commercial communication more effective in each context in which it appears.

For instance, in a TV programs largely attended by male persons (such as a football match), the spot could be broadcasted in the "male" version and vice versa for a program largely attended by a female audience.

From the results of the biometric recordings obtained for the men and women groups, the proposed time reduction of the ad is presented follows in the next Figure (Figure 52).

For men, the seconds of the spot concerning the description of the high technical characteristics of the car were kept and privileged; whereas, for women, frames relative to features of safety and comfort were selected, enhancing the presence of the female Figure and highlighting the three little girls, just as emerged from the neurometric recordings.

In both cases the content and the advertising messages of the spot have been respected, but the commercial has been customized to the emotions as aroused in male and female target.

Fig. 52 The time reduction proposed for the sub-targets: men and women. On the left, the frames related to the commercial reduction by taking into account the information conveyed by the male emotional profile. On the right, the frames related to the commercial reduction by taking into account the female emotional profile. It could be observed as two distinct spots related to the Giulietta material have been derived from the application of the neurometric indexes in this case.

9.8 Analysis of the Impact of a TV Commercial in the Insurance Sector on Two Different Categories of Users: “Agency” and “On line”

In this chapter another example of the neuroelectrical tools application is presented relative to the evaluation of a spot efficacy on a TV commercial for the car insurance sector. The study was carried out in Italy on October 2012 on a sample of 24 men, aged between 34 and 54. One of the goals of the neurometric test was to obtain useful insights to increase the impact of the communication strategy toward specific groups of persons potentially interested in such type of communication. In particular, preliminary interviews have highlighted the existence of two subgroups of persons: 1) a first group (called the “Agency” group) is formed by persons that have a car insurance contract with a traditional agency, (e.g. traditional office, located in a street etc.); 2) a second group is formed by persons that have their car insurance contract obtained by an on-line agency (the “On-line” group).

The tested commercial spot is devoted to advertise a new company that allows the comparisons of the car insurance offering among different on-line agencies on the WEB. Thus, it was supposed to produce different effects on such different categories of potential users (“Agency” and “On-line”). In fact, subjects having an insurance contract in agency (who are less familiar with digital devices and probably more diffident about on-line insurance) were compared to their counterpart subscribing on-line insurance contract, more familiar with technology devices.

In addition to Attention, Memorization and Emotional indexes, already used in the already presented examples, in this chapter another biometric indicator of cognitive interest for a sensory percept has been used. Such index, called "Interest", is based on a neuroscience theory published since the 1990s by the research group of Davidson and coworkers. Briefly, it has been demonstrated, by a lot of published research, as in our brain two main prefrontal cortical systems regulate our tendency to approach or to withdraw from perceived stimuli. In particular, the Behavioral Approach System (BAS) is guiding us toward appreciated stimuli (e.g. a cold drink during summer) while another prefrontal system (the Behavioral Inhibitory System; BIS) is pushing us to withdraw when we perceived unpleasant or not useful sensory stimuli. Such BIS and BAS prefrontal activations could be indexed by a cerebral index, here called "Interest" and in other publications from other Authors is called "likeness". Here it has been preferred the name "interest" since a measure of the emotional likeness provided by the body markers though heart rate (HR) and Galvanic Skin Response (GSR) was already used for the emotional index.

Thus, values of the "Interest" indicator oscillates between positive values, indicating a prevalence of the BAS prefrontal activations (suggesting interest for the user), to negative values of it, relate to a prevalence of the BIS prefrontal activations for the target sample (suggesting low interest for the group).

The neurometric indexes for Memorization, Attention, and Interest used in this example have been evaluated using z-scores values. The temporal evolution of the tested TV commercial was divided in different time segments, relative to particular portion of the investigated commercial. In particular, one time segment was called *Introduction* collecting all the frames of the commercial that was initially offered to both groups. Successively, it was considered the *Service* temporal segment, in which the particular service offered by the company is presented. The *Brand* segment was instead relative to the presentation of the brand of the company that promoted the advertisement, while the *All* segment is relative to the entire spot.

Average values of the neurometric indexes employed in this study were computed in each considered segment for the two subgroup populations ("Agency" and "On-line"). Such indexes values are reported in Figure 53. In such Figure, it is possible to note how the tested spot generates different values of the neurometric indexes for the two analyzed groups.

The "Agency" group presented values of the memorization index better than those obtained in the "On-line" group in all the considered time segments. Also the emotional response was positive for the "Agency" group and better in all segments but "Introduction" when compared with the "On-line" group.

On the other hand, the "On-line" group presented higher values of interest and attention indexes in all segments with respect to the "Agency" group.

A possible interpretation of these findings is that the "Agency" group has been more emotionally involved on the "Service" segment, showing the opportunity of saving costs by using on-line insurance (that they have not). In addition, this increase of emotion of the "Agency" group also happens when the brand appeared on the screen since the possibility to save money with on-line services was new to them. In fact, the "On-line" group already was informed of this possibility since already owns on-line insurances.

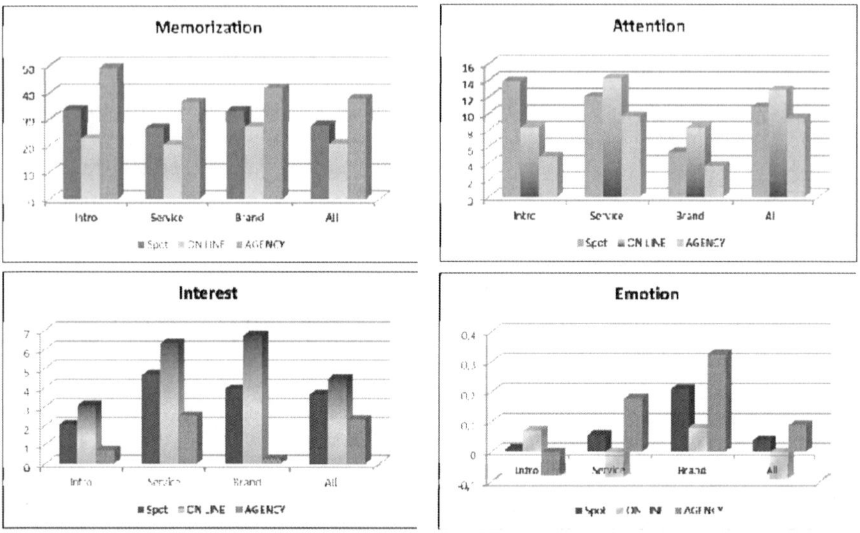

Fig. 53 Average evaluation on different spot segments (Intro, Offer of the Service, Brand, Whole spot) of cognitive and emotional indexes. For each segment, the average is evaluated in the whole sample (first column) and in two sub- targets: "auto insurance buyers on line" (second column) and "auto insurance buyers in agency" (third column).

As described above, the results of the neurometric test showed in Figure 53 suggest how the "Agency" group has less interest in the Service and Brand section of the spot when compared to the "On-line" group. A possible interpretation of this finding is related to the low tendency of the "Agency" group to retain "technical" elements that could have the attempt to shift the comprehension of the analytic element of the offer later. On the other hand, the "On-line" group presented higher values of interest and attention towards the spot in many of the time segments that offered comparative and economic information to the user. A possible interpretation of this finding is that the "On-line" group is composed by persons who already know the service and has interest to know more in detail about the advantage of the service.

Likely, they want to capture more details from the offer, probably in order to understand if the service is more advantageous for them. Thus, it could be important for the "On-line" group giving clear presentation of the convenience features of the offer.

The "Agency" group just enjoyed the fact that a saving is possible, but the general low technical competence, when compared to the other group, suggest them to shift the cognitive comprehension of the service to a later moment (maybe asking a friend or other actions).

The analysis of the differences between the sub-targets may be significant to emphasize key insights in communication strategy. In this case both groups seemed needing of enhancing corporate branding for a better recall of the company. Moreover, the neurometric indexes suggested to take into account a better highlight of the "saving offer" in the general communication. Another element arose from the analysis of neurometric indexes that is to simplify the message of the proposal, by shifting the technical detail of the services on the successive WEB page. This could make less hard, for non-technical people, to understand the benefit of the service.

The provided example shows as the use of neurometric methodologies may be helpful not only for performing analysis of the efficacy (second by second) of a TV commercial, but also to discover insights of the person's spot perception, which is difficult to achieve with traditional research tools.

9.9 Cultural and Local Differences

The aim of this chapter is to analyze if the spot, previously shown, also presented differences in the perception between other two sub-targets derived by the same sample. They are composed by subjects living in Milan (northern Italy) and subjects living in Rome (central Italy). Same indexes as in the previous chapter were used. The results are presented in Figure 54:

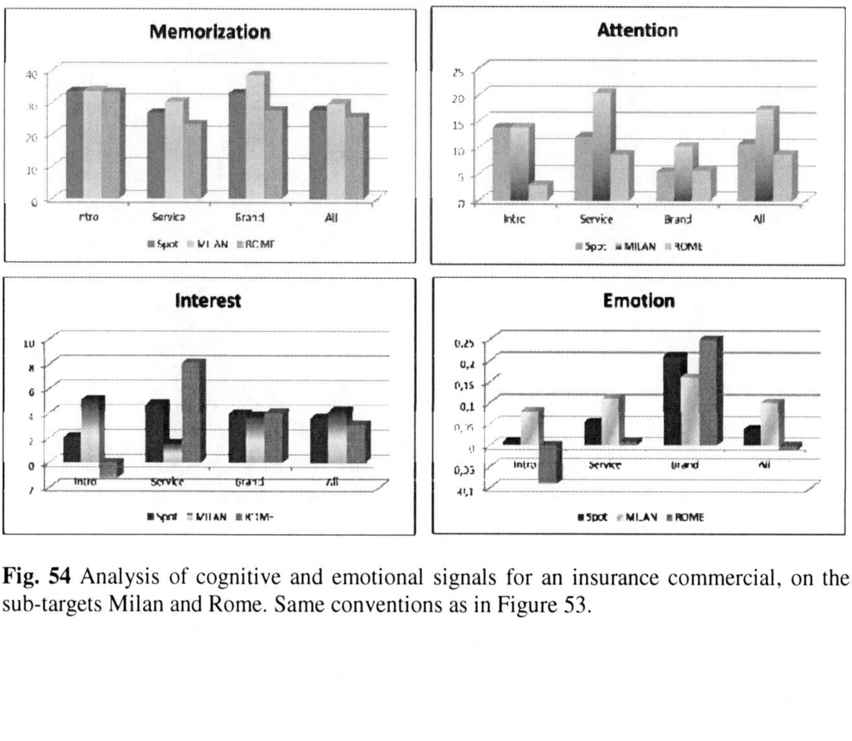

Fig. 54 Analysis of cognitive and emotional signals for an insurance commercial, on the sub-targets Milan and Rome. Same conventions as in Figure 53.

- In Milan, testers presented higher level of memorization and attention indexes than people in Rome, particularly for "Service" and "Brand" segments; also emotion index is higher in all segments except "Brand".
- In Rome, the testers presented higher values of interest than in Milan as to the "Service" segment, while the emotional index is higher than in Milan only in the "Brand" segment.
- In Milan, people presented higher values of the memorization and attention indexes during the entire spot when compared to the group in Rome. Subjects living in Milan are also presenting higher values of the emotional indexes than the subjects living in Rome.
- In Rome instead, people were more interested in the "service" segment and more emotionally involved by the "brand" segment. The emotional level was always lower in the other segments of the spot.

The observed differences in the neurometric average values, for the analyzed subgroups suggested to better qualitatively investigate the different way of thinking and perceiving the creative solution and the offer for the groups of persons living in Rome and Milan. Furthermore, from these results the client was recommended to deeply listening people from these two towns by qualitative research in order to better understand eventual different insights and possible improvement in the future of the efficacy of communication by using specific local advertisements.

Chapter 10
The Next Frontier in Neuroimaging Methodologies: Assessing the Functional Connections between Cortical Areas

In the previous chapters it has been described the methodological approaches useful for the analysis of neurophysiological data provided by the EEG recordings related to marketing stimuli. Such approaches involved essentially the estimation of the power spectrum of the EEG data in particular cortical locations, such as the prefrontal and orbitofrontal cortices, estimated by using the solution of the electromagnetic linear inverse problem (as described in the chapter 2). In addition, statistical analysis of the resulting cortical estimated distributions have also illustrated in order to get robust results from the undergoing EEG "noise" (e.g. EEG activity not related to the observation of marketing stimuli). However, the possible methodologies to be applied for extract meaningful information for the brain processing occurring during observation of advertising stimuli are not limited to such brain imaging procedures. In fact, the proposed methodologies are particularly useful for a detection of the main psychological attitudes of the subjects in front of marketing stimuli, such as memorization, attention or even emotion. Such approaches are well grounded and supported by several published studies in scientific literature (some also cited in this book) that assures the applicability of the presented methodology. This last statement is particularly important since in the field of the application of neurophysiological measurements to the field of evaluation of marketing stimuli there are many unsupported methodologies and claims from private companies that have not published their methodologies on scientific journals.

Other advanced methodologies when compared to the estimation of cortical activity are currently under development in order to extract more information on the brain data related to audiovisual stimuli. Surely, one of the most interesting methodology in this area is the estimation of the degree of cooperation between the different cortical areas in the analysis of the received stimuli. Such degree of cooperation could be established by using EEG or even hemodynamic recordings and particular computational techniques we are going to illustrate in the next paragraphs. In particular, it will be explored the idea of the covariance paradigm,

G. Vecchiato et al.: *Neuroelectrical Brain Imaging Tools*, BIOSYSROB 3, pp. 113 – 125.
DOI: 10.1007/978-3-642-38064-8_10

that considers of interest the evaluation of the statistical robustness of the computational links between the cerebral activities detected in different brain areas. The assertion here is that the task represented by an experimental condition is mediated by a network of interacting brain regions, and that different tasks correspond to different functional networks. Thus, by examining the covariance in brain activity between different brain areas, one can infer something about which areas are important nodes in the network under study, and how these nodes are functionally connected.

10.1 Connectivity Modeling (Structural-Functional-Effective)

The term brain connectivity refers to several different and interrelated aspects of brain organization (Horwitz, 2003). A fundamental distinction is that between structural connectivity, functional connectivity and effective connectivity (Friston, 1994). Structural connectivity refers to a network of physical (synaptic) connections linking sets of neurons or neuronal elements, as well as their associated structural biophysical attributes encapsulated in parameters such as synaptic strength or effectiveness. Structural connectivity is relatively stable at short time scales (seconds to minutes), but at longer time scales (hours to days), it is likely to be subject to significant morphological change and plasticity.

Functional connectivity describes statistical dependence between distributed and often spatially remote neuronal units. It can be estimated in a variety of ways, for example through computing cross-correlations in the time or frequency domain, mutual information or spectral coherence. Whereas functional connectivity captures patterns of statistical dependence, effective connectivity describes networks of causal influences of one brain area over another. Various techniques for extracting effective connectivity have been pursued. One technique called "covariance structural equation modelling" assigns effective connection strengths to anatomical pathways that best match observed covariances in a given task (Mclntosh and Gonzalez-Lima, 1994). A generalization of this approach called "dynamic causal modelling" (Friston et al., 2003) operates in a Bayesian framework to estimate and make inferences about directed influences between variables. Yet another approach to identifying highly interactive brain regions and their directional interactions involves the use of effective information, a measure that uses a perturbation approach to capture the degree to which two brain regions or systems causally influence each other (Tononi and Sporns, 2003). Effective connectivity may also be estimated on the basis of time-series analysis. Some of these methods are based on interpretations or adaptations of the concept of Granger causality. However, the estimation of possible links between the brain activity data collected is not limited to the neuroelectrical recordings (EEG or MEG), but could be also be extended to the evaluation of the hemodynamical recordings as furnished by using the functional magnetic resonance imaging (fMRI).

10.2 A Connectivity Modeling Based on Hemodynamic Brain Data: The Structural Equation Modeling (SEM)

In Structural Equation Modeling (SEM), the parameters are estimated by minimizing the difference between the observed covariances and those implied by a structural or path model. In terms of neural systems, a measure of covariance represents the degree to which the activities of two or more regions are related. The SEM consists of a set of linear structural equations containing observed variables and parameters defining causal relationships among the variables. Variables in the equation system can be endogenous (i.e. dependent from the other variables in the model) or exogenous (independent from the model itself). The structural equation model specifies the causal relationship among the variables, describes the causal effects and assigns the explained and the unexplained variance. Let us consider a set of variables (expressed as deviations from their means) with N observations. In the context in which this technique is introduced, these variables represent the activity estimated in each cortical region, during the exposition to a particular marketing message

The SEM for such variables is the following:

$$\mathbf{y} = \mathbf{B}\mathbf{y} + \boldsymbol{\Gamma}\mathbf{x} + \zeta \tag{38}$$

where:

- $\mathbf{y}$ is a (m x 1) vector of dependent (endogenous) variables;
- $\mathbf{x}$ is a (n x 1) vector of independent (exogenous) variables;
- ζ is a (m x 1) vector of equation errors (random disturbances);
- $\mathbf{B}$ is a (m x m) matrix of coefficients of the endogenous variables;
- $\boldsymbol{\Gamma}$ is a (m x n) matrix of coefficients of the exogenous variables.

ζ is assumed to be uncorrelated with the exogenous variables, and B is supposed to have zeros in its diagonal (i.e., an endogenous variable does not influence itself) and to satisfy the assumption that (I–B) is non-singular, where I is the identity matrix.

The covariance matrices of this model are the following:

- $\boldsymbol{\Phi} = E[\mathbf{x}\,\mathbf{x}^T]$ is the (n×n) covariance matrix of the exogenous variables;
- $\boldsymbol{\Psi} = E[\zeta\,\zeta^T]$ is the (m×m) covariance matrix of the errors.

If $\mathbf{z}$ is a vector containing all the p=m+n variables, exogenous and endogenous, in the following order:

$$\mathbf{z}^T = [x_1 \ldots x_n\, y_1 \ldots y_m] \tag{39}$$

the observed covariances can be expressed as:

$$\boldsymbol{\Sigma}_{obs} = (1/(N\text{-}1))\cdot\mathbf{Z}\cdot\mathbf{Z}^T \tag{40}$$

where $\mathbf{Z}$ is the $p{\times}N$ matrix of the p observed variables for N observations.

The covariance matrix implied by the model can be obtained as follows:

$$\mathbf{\Sigma}_{\text{mod}} = E[\mathbf{z}^T\mathbf{z}] = \begin{bmatrix} E[\mathbf{x}\mathbf{x}^T] & E[\mathbf{x}\mathbf{y}^T] \\ E[\mathbf{y}\mathbf{x}^T] & E[\mathbf{y}\mathbf{y}^T] \end{bmatrix} \tag{41}$$

where:

$$\begin{aligned} \mathrm{E}[\mathbf{y}\mathbf{y}^{\mathrm{T}}] &= \mathrm{E}[(\mathbf{I}\text{-}\mathbf{B})^{-1}(\mathbf{\Gamma}\mathbf{x} + \zeta)(\mathbf{\Gamma}\mathbf{x} + \zeta)^{\mathrm{T}}((\mathbf{I}\text{-}\mathbf{B})^{-1})^{\mathrm{T}}] \\ &= (\mathbf{I}\text{-}\mathbf{B})^{-1}(\mathbf{\Gamma}\mathbf{\Phi}\mathbf{\Gamma} + \mathbf{\Psi})((\mathbf{I}\text{-}\mathbf{B})^{-1})^{\mathrm{T}} \end{aligned} \tag{42}$$

since the errors ζ are not correlated with the x;

$$\mathrm{E}[\mathbf{x}\mathbf{x}^{\mathrm{T}}] = \mathbf{\Phi} \tag{43}$$

$$\mathrm{E}[\mathbf{x}\mathbf{y}^{\mathrm{T}}] = (\mathbf{I}\text{-}\mathbf{B})^{-1}\,\mathbf{\Phi} \tag{44}$$

$$\mathrm{E}[\mathbf{y}\mathbf{x}^{\mathrm{T}}] = ((\mathbf{I}\text{-}\mathbf{B})^{-1}\,\mathbf{\Phi})^{\mathrm{T}} \tag{45}$$

since $\mathbf{\Sigma}_{\text{mod}}$ is symmetric. The resulting covariance matrix, in terms of the model parameters, is the following:

$$\mathbf{\Sigma}_{\text{mod}} = \begin{bmatrix} \mathbf{\Phi} & (\mathbf{I}-\mathbf{B})^{-1}\mathbf{\Phi} \\ ((\mathbf{I}-\mathbf{B})^{-1}\mathbf{\Phi})^T & (\mathbf{I}-\mathbf{B})^{-1}(\mathbf{\Gamma}\mathbf{\Phi}\mathbf{\Gamma}+\mathbf{\Psi})((\mathbf{I}-\mathbf{B})^{-1})^T \end{bmatrix} \tag{46}$$

Without other constraints, the problem of the minimization of the differences between the observed covariance and those implied by the model is underdetermined, because the number of variables (elements of matrices $\mathbf{B}$, $\mathbf{\Gamma}$, $\mathbf{\Psi}$ and $\mathbf{\Phi}$) is greater than the number of equations (m+n)(m+n+1)/2. For this reason, the SEM technique is based on the *a priori* formulation of a model, on the basis of anatomical and physiological constraints. This model implies the existence of just some causal relationships among variables, represented by arcs in a "path" diagram; all the parameters related to arcs not present in the hypothesized model are forced to zero. For this reason, all the parameters to be estimated are called free parameters. If t is the number of free parameters, it must be t<=(m+n)(m+n+1)/2.

These parameters are estimated by minimizing a function of the observed and implied covariance matrices. The most widely used objective function for SEM is the maximum likelihood (ML) function:

$$F_{ML} = \log|\mathbf{\Sigma}_{mod}| + \mathrm{tr}(\mathbf{\Sigma}_{obs}\cdot\mathbf{\Sigma}_{mod}^{-1}) - \log|\mathbf{\Sigma}_{obs}| - p \tag{47}$$

where tr(·) is the trace of matrix. In the context of multivariate, normally distributed variables the minimum of the maximum likelihood function, multiplied by (N-1), follows a χ^2 distribution with $p(p + 1)/2 - t$ degrees of freedom, where t is the number of parameters to be estimated and p is the total number of observed variables (endogenous+exogenous). The χ^2 statistic test can then be used to infer statistical significance of the structural equation model obtained. The public available software LISREL could be used for the implementation of the SEM technique.

SEM technique is particularly useful when a predefined model of the interaction between the cortical areas is available. Although SEM technique could be also adopted by using neuroelectromagnetic data (EEG or MEG), most applications in literature of such modeling were performed with the use of hemodynamic data. When no a priori information are available to model the interactions between the cortical areas other approached could be attempted. Such approaches will be presented in the next paragraphs.

10.3 Estimating Functional Connections between Brain Areas by Using Neuroelectromagnetic Data

The evaluation of a connectivity modeling by using EEG or MEG data applied to the marketing stimuli could be performed by using one of the different methodologies available to estimate directional coupling. One of the available methodology is the Directed Transfer Function technique (DTF, Kaminski and Blinowska, 1991). However, also other directional connectivity estimation could be used to estimate functional connectivity between cortical areas, such as the Partial Directed Coherence for instance (Baccalà and Sameshima, 2001). In general, the DTF could be applied on a set of cortical estimated waveforms Z as suggested by the methodologies illustrated in the chapter 2 of this book as in the following. Let **Z**(t) the vector of the estimated cortical signals :

$$\mathbf{Z}(t) = [z_1(t), z_2(t), \ldots, z_N(t)]^T \tag{48}$$

The following MVAR process is an adequate description of the data set Z:

$$\sum_{k=0}^{q} \mathbf{\Lambda}(k)\mathbf{z}(t-k) = \mathbf{e}(t) \quad , \text{ with } \mathbf{\Lambda}(0) = \mathbf{I} \tag{49}$$

where **e**(t) is a vector of multivariate zero-mean uncorrelated white noise process, $\mathbf{\Lambda}(1)$, $\mathbf{\Lambda}(2)$, … $\mathbf{\Lambda}(q)$ are the NxN matrices of model coefficients and q is the model order, chosen, for example, with the Akaike Information Criteria for MVAR process (Kaminski et al., 2001). In order to investigate the spectral properties of the examined process, the Eq. (49) is transformed to the frequency domain:

$$\mathbf{\Lambda}(f)\, \mathbf{Z}(f) = \mathbf{E}(f) \tag{50}$$

where:

$$\mathbf{\Lambda}(f) = \sum_{k=0}^{q} \mathbf{\Lambda}(k) e^{-j2\pi f \Delta t k} \tag{51}$$

and Δt is the temporal interval between two samples. Eq. (50) can then be rewritten as:

$$\mathbf{Z}(f) = \mathbf{\Lambda}^{-1}(f)\,\mathbf{E}(f) = \mathbf{H}(f)\,\mathbf{E}(f) \tag{52}$$

H(f) is the transfer matrix of the system, whose element H_{ij} represents the connection between the j-th estimated cortical brain signal and the i-th signal of the system. With these definitions, the causal influence of the cortical waveform estimated in the j-th brain area on that estimated in the i-th brain area (the Directed Transfer Function $\theta^2_{ij}(f)$) is defined as:

$$\theta^2_{ij}(f) = \left|H_{ij}(f)\right|^2 \tag{53}$$

In order to be able to compare the results obtained for cortical waveforms with different power spectra, a normalization was performed by dividing each estimated DTF by the squared sums of all elements of the relevant row, thus obtaining the so called normalized DTF (Kaminski and Blinowska, 1991):

$$\gamma^2_{ij}(f) = \frac{\left|H_{ij}(f)\right|^2}{\sum_{m=1}^{N}\left|H_{im}(f)\right|^2} \tag{54}$$

The $\gamma_{ij}(f)$ expresses the ratio of influence of the cortical waveform estimated in the j-th brain area on the cortical waveform estimated on the i-th brain area, with respect to the influence of all the estimated cortical waveforms. Normalized DTF values are obtained when the normalization condition

$$\sum_{n=1}^{N}\gamma^2_{in}(f) = 1 \tag{55}$$

is applied. One obstacle in the application of the DTF to the estimation of cortical activity during the observation of visual and auditory stimuli is related to the level of statistical significance that the estimated connections must reach in order to be considered different from the background noise. In this case a bootstrapping or a shuffling procedure could be recommended (Kaminski et al., 2001).

While the estimation of spectral properties for the EEG is nowadays performed also in real time, on-line estimation of the cortical and scalp connectivity cannot be performed nowadays with the standard computational resources available. However, with the doubling of the computational power every few years such limitation could be removed within the next five years.

10.4 Beyond the Estimation of Cortical Connectivity: Time-Invariant Graph Theory Measures

As described before, irrespective from the neuroimaging methodology employed, by using hemodynamic or neuro-electric waveforms, it is always possible to estimate a network of statistical significant connectivity values (by DTF, PDC or SEM) that linked functionally different scalp or cortical areas. However, the information in the resulting cortical connectivity networks could be difficult to be evaluated by eyes. Thus, a further step in methoodlogy is required, by using computational approaches and tools belonging to the graph theory.

The theory of directed graphs is of special interest as it applies to structural, functional and effective brain connectivity at all levels. Since a graph is a mathematical representation of a network, which is essentially reduced to nodes and connections between them, a way to characterize topographical properties of real complex networks was proposed using a graph theoretical approach (Sporns and Zwj, 2004; Stam, 2004; Stam et al., 2006). It was realized that functional connectivity networks estimated from EEG, MEG or fMRI recordings can be analyzed with tools that have been already generated for the treatments of graphs as mathematical objects (Sporns and Zwj, 2004). This is interesting since the use of mathematical indexes for summarizing some graph properties allows for the generation and testing of particular hypotheses on the physiologic nature of the functional networks estimated from high-resolution EEG recordings. The use of graph tools seem to be particularly adequate to characterize functional connectivity patterns estimated from high-resolution EEG or MEG data. Watts and Strogatz have shown that graphs with many local connections and few random long distance connections are characterized by a high cluster index C and a short path length L (Watts and Strogatz; 1998). Such near optimal models are designated as "small-world" networks. Many types of real networks have been shown to share these small-world features (Boccaletti, 2006; Strogaz, 2001, Latora et al., 2001). Patterns of anatomical connectivity in neuronal networks are particularly characterized by high clustering and a small path length. Networks of functional connectivity based upon fMRI BOLD signals or MEG recordings have also been shown to have small-world features (Boccaletti, 2006; Strogaz, 2001, Latora et al., 2001).

10.4.1 Scale-Free Networks

Besides, some real networks are mostly found to be very unlike the random graph in their degree distributions. It was demonstrated that such degree distributions follow a power law trend (Barabasi et al.; 1999). Those networks, called "scale-free", also exhibit the small-world phenomenon, but tend to contain few nodes that act as highly connected "hubs", although most of the nodes have low degrees. Scale-free networks are very peculiar in how they respond to damages. An interesting characteristic of such networks is that they are extremely tolerant of random failures. In fact they can absorb random failures up to about 75% of their nodes before they collapse, while they are more vulnerable to intentional attacks on their hubs. It is important then understand how the real physiologic networks estimated in this analysis from EEG, MEG data can be characterized.

10.4.2 Degrees and Distributions

The simpler attribute for a graph's node is its degree of connectivity, which is the total number of connections with other nodes. The arithmetical average of all nodes' degree is called mean degree of the graph. Indeed, this mean value gives little information about the behavior of degree within the system. Hence, it is useful to introduce P(k) as the fraction of vertices in the graph that have degree k. Equivalently, P(k) is the probability that a vertex chosen uniformly at random has degree k. A plot of P(k) for any given network can be constructed by making a histogram of the degrees of vertices. This histogram is the degree distribution for the graph and it allows to better understand the degree allocation in the system. Here, because the graph obtained from EEG and MEG data will be directed, the in-degree kin and the out-degree kout for each node must be considered separately. They represent the total number of connections incoming (afferent) to a vertex and outgoing (efferent) from the same vertex, respectively. Degrees have obvious functional interpretations. A high degree-in indicates that a neural region is influenced by a large number of other areas, while a high degree-out indicates a large number of potential functional targets.

10.4.3 Efficiency

The efficiency is a quantity recently introduced in (Latora and Marchiori, 2001) to measure how efficiently the nodes of the network communicate if they exchange information in parallel. It is computed from the distance matrix D, that contains distances between each pair of nodes.

The efficiency eij in the communication between vertices i and j can then be defined to be inversely proportional to the shortest distance:

$$eij = 1/dij \tag{56}$$

When there is no path in the graph between i and j, dij = ∞; and consistently, eij = 0. Global efficiency of A can be defined as:

$$Eglob = 1/ (N * N-1) * \Sigma j\ (1/dj) \tag{57}$$

where N is the number of vertices composing the graph. Note that 1/L (inverse of the characteristic path length) can be seen as first approximation of Eglob. Since the efficiency is also defined for disconnected graphs the local properties of A can be characterized by evaluating for each vertex i the efficiency of Ai, the sub-graph of the first neighbors of i. Local efficiency (Eloc) is the average of all the sub-graphs' global efficiencies. The local efficiency Eloc reveals how much the system is fault tolerant, thus it shows how efficient the communication is among the neighbors of i when i is removed. Connection matrix contains values for each directed pair of cortical areas considered and can be converted to an adjacency matrix A by considering a threshold T that represents the number of the most powerful arcs to be considered. If the number of links in the connection matrix exceeds T, less powerful edges will be removed until the reaching of that threshold. Also, T can be expressed as connection density that is the ratio between the number of all the effective connections and the number of all possible connections within the graph. In order to study the topology of networks at different connection densities or costs (Latora and Marchiori, 2003) a range of threshold values T will be generally explored.

Many scientific studies have shown that the cerebral networks exhibited a small-world architecture during no-tasks and/or rest periods (Salvador et al., 2005; Achard and Bullmore, 2007; Stam, 2004). In an application to the evaluation of marketing stimuli (Fallani et al., 2008) , the small-world characteristic was found during a period of time in which an unsuccessful memorization of TV commercials is performed by the analyzed population. Besides the local properties, also the global-efficiency index showed a significant decrease between the TV commercials that have been forgotten (FRG dataset) and those that have been remembered (RMB dataset) in the beta and gamma frequency bands. In such study, the estimated cortical networks returned further information about the complex interactions among the considered cortical areas. In fact, during the RMB situation, the functional network in the beta and gamma band presented a significant non-homogeneous allocation of the involved information flows and a consequent reduction of the efficiency in the overall communication between the network nodes. In the beta and gamma frequency bands, the respective reduction of global-efficiency as well as the reduction of local-efficiency for the alpha band of the cortical network communication could represent a predictive measure for the accurate recall of the commercials that will be remembered. On the base of the obtained experimental results from the application of theoretical graph indexes to the functional networks estimated by using advanced high-resolution EEG, Authors observed three phenomena linked to the functional re-organization of the cortical connectivity networks during the observation of TV commercials. In particular they observed that:

(1) The structure of cortical network presented significant differences during the visualization of the TV spots that were remembered with respect to those that were forgotten by the analyzed population. In particular, the successful encoding of TV spots produced an evident decrease of the average values of global- and local-efficiency in the estimated cortical networks.
(2) The main differences between the two analyzed conditions (FRG and RMB) appeared to be in the beta and gamma bands for the global-efficiency index and in the alpha band for the local-efficiency index.
(3) In the light of the comparison performed with random graphs, the optimal level of global and local communication – i.e. small-world network – during

the visualization of the spots that will not be remembered (FRG) reflects a low level of neural involvement. Instead, the presence of attentive and semantic processes during the visualization of the video-clips that will be remembered (RMB) leads to a weak pattern of communication among the considered cortical areas analyzed. These considerations suggest the possibility to use in a next future such sophisticated technologies to predict the memorization ability of the observation of TV commercials.

10.5 Beyond the Estimation of Cortical Connectivity: Time-Varying Graph Theory Measures

Graphs of brain networks can be quantitatively examined for vertex degrees and strengths, degree correlations, sub-graphs (motifs), clustering coefficients, path lengths (distances), and vertex and edge centrality, among many other graph theory measures. The statistical evaluation of these measures requires the design of appropriate null hypotheses, and this will be an active area for further research. Clustering techniques and quantitative measures of dynamic changes will employ a range of new clustering algorithms some based on principal components analysis, multidimensional scaling and the graph's eigenspectrum (Fallani et al., 2011).

One of the goals of the application of the graph theory to the analysis of the data related to the brain activity in the observation of marketing stimuli is to derive new set of features to better describe the memorization and attention cerebral activities during the fruition of marketing stimuli. One possible idea, still not applicable due to the limitation of the computational power available for the normal computer nowadays is to estimate time-varying cortical network indices to be linked with the fruition of such marketing stimuli. It is important to note that such indexes could return properties of the different analyzed networks that can be subjected to multivariate statistical analysis and classification. Consider a sequence of temporal lags l = 1, 2, ..., L (which can represent second or minutes during the exposition to the marketing stimuli), the functional connectivity between N brain signals $X = [x_1, x_2, ..., x_N]$ can be represented at each lag l with a connectivity matrix C (N x N):

$$C = \begin{pmatrix} c(1,1) & \cdots & c(N,1) \\ \vdots & \ddots & \vdots \\ c(1,N) & \cdots & c(N,N) \end{pmatrix} \tag{58}$$

where the element c(i,j) with I = 1, …, N and j = 1, …, N, is a measure of the information flow from the region i to j. The topology of the connectivity pattern at each time lag l can be characterized by a set of K parameters $g_1, g_2, ..., g_K$, each one extracted from the connectivity matrices C. Such a set of parameters can provide a parametrization of the changes in connectivity across each time lag considered (Figure 55).

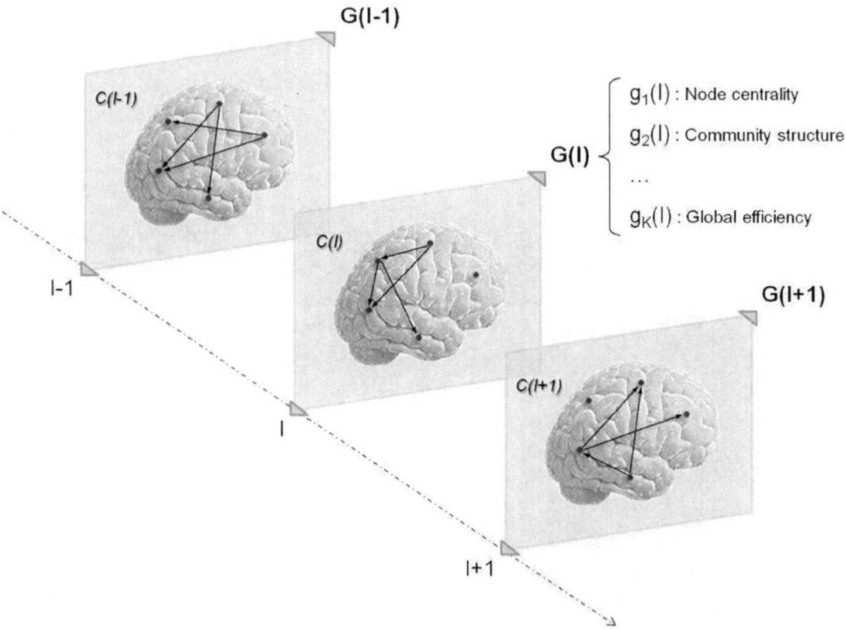

Fig. 55 Schematic of a spatio-temporal functional brain network. Only three instants are represented for the sake of simplicity i.e. l-1, l, l+1. At each instant, K.

The multivariate variable G(l) contains information about the time evolution of different spatial network properties. Notably, a multivariate matrix G (K x L) can be obtained as:

$$G = \begin{pmatrix} g_1(1) & g_1(2) & \cdots & g_1(L) \\ \vdots & & & \vdots \\ g_k(1) & g_k(2) & \cdots & g_k(L) \end{pmatrix} \tag{59}$$

Such characteristic matrix provides one possible approach spatio-temporal features of the functional brain networks. Actually, each row of the matrix can be thought as a new signal of L samples describing the evolution of a particular network parameter. From this G matrix, mean, variance, as well as possible higher-order statistics, can be derived to assess linear/non linear relations within and between the temporal evolutions of the different network indices. The formalism is well suited for predicting the connectivity state of a given brain area or network, for example through regression analysis. Obviously, a data-driven approach that allows knowledge of the past brain network properties to reliably predict future functional organization is especially relevant for any attempt to optimize the delivery of the proposed marketing process.

10.6 Concluding Remarks

In the previous paragraphs of this chapter more advanced approaches with respect to the evaluation of the spectral power of the EEG signals in predefined cortical areas have been presented. Such approaches involved the concept of functional connectivity in the analysis of time series obtained from the brain activity. The concept of functional connectivity is what is actually believed to be the key to a further and deep understanding of brain behavior in response to complicate external world stimulation, as for instance an audiovisual stimulation. In fact, nowadays it is possible to obtain with the computational tools already available (such as DTF or SEM, but also with the PDC and many others) an estimation of the cortical network subserving a particular cognitive task experienced or executed by the subject. Such estimated cortical network could be actually obtained off-line, i.e. not in real time when compared to the experiment performance by the subjects, due to the particular computational load required by such kind of analysis. In fact, actually the time-varying computation of functional connectivity links, as obtained by time varying version of PDC, DTF and other methodologies are actually quite computationally cumbersome. The same is true for the application of the computational tools available for the characterization of the estimated network properties by using graph theory. Also in this respect, the applicability of global and local efficiency of a graph in real time with the execution of the subject's task is again a dream today. Many researchers are actually working refining the algorithms available for the estimation of cortical activity and connectivity in order to have soon reliable methodologies ready to be employed on-line. These future algorithms will allow the incorporation of such computational tools in devices that will enlarge the application of bioengineering methodologies in different fields of science and technology.

In the marketing area, the availability of more advanced and refined signal processing algorithms will make easier the generation of reliable results related to the brain signal analysis and interpretation during the observation of commercial advertisings in real time. In fact, it is true that all the results here reported have a time delay between the actual neuroelectromagnetic recordings and the formulation of the results of about 5-8 days. More advanced signal processing algorithms could promote the availability of such results if not on-line on very few days (2-3) after the recordings. This could help a lot the industries in the refinement of their advertisings without to waste time and money due to an incorrect broadcasting of a incorrect or "wrong" TV commercials.

Allowing the industries to save their money from the potential waste due to bad TV commercials means that more money could be instead employed to promote the research for their product, improving at the end the quality of the products and the final benefit for the users.

Marketing strategies are usually employed to generate products more close and useful to the users and not viceversa (i.e convincing the users to get unuseful products from industries). We think that this point is often overlooked by many scientists with the criticisms to the application of neuroscience and engineering techniques to the marketing methodologies. Engineering sciences are usually called to give robust and scientifically-based answers to the needs of the society, and today also in the marketing science. As pointed out several centuries ago by the scientists and artist Leonardo da Vinci (1452-1519), "Who loves practice without theory is like the sailor who boards ship without a rudder and compass and never knows where he may cast".

References

Adolphs, R.: Cognitive neuroscience of human social behaviour. Nature Reviews. Neuroscience 4(3), 165–178 (2003)

Aftanas, L.I., Reva, N.V., Varlamov, A.A., Pavlov, S.V., Makhnev, V.P.: Analysis of evoked EEG synchronization and desynchronization in conditions of emotional activation in humans: temporal and topographic characteristics. Neuroscience and Behavioral Physiology 34(8), 859–867 (2004)

Aharon, I., et al.: Beautiful faces have variable reward value: fMRI and behavioral evidence. Neuron 32(3), 537–551 (2001)

Ambler, T., Burne, T.: The Impact of Affect on Memory of Advertising. Journal of Advertising Research 39(2), 25–34 (1999)

Ambler, T., Ioannides, A., Rose, S.: Brands on the Brain: Neuro-Images of Advertising. Business Strategy Review 11(3), 17–30 (2000)

Anderson, S.J., Glantz, S., Ling, P.: Emotions for sale: Cigarette advertising and women's psychosocial needs. Tobacco Control 14, 127–135 (2005)

Ariely, D., Berns: Neuromarketing: the hope and hype of neuroimaging in business. Nature Reviews. Neuroscience 11(4), 284–292 (2010)

Arnow, B.A., et al.: Brain activation and sexual arousal in healthy, heterosexual males. Brain: A Journal of Neurology 125(pt. 5), 1014–1023 (2002)

Astolfi, L., Vecchiato, G., De Vico Fallani, F., Salinari, S., Cincotti, F., Aloise, F., Mattia, D., et al.: The track of brain activ-ity during the observation of TV commercials with the high-resolution EEG technology. Computational Intelligence and Neuroscience, 652078 (2009)

Astolfi, L., et al.: Tracking the Time-Varying Cortical Connectivity Patterns by Adaptive Multivariate Estimators. IEEE Transactions on Biomedical Engineering 55(3), 902–913 (2008a)

Astolfi, L., et al.: Brain activity related to the memorization of TV commercials. International Journal of Bioelectromegnetism 10(2), 1–10 (2008b)

Astolfi, L., et al.: Neural basis for brain responses to TV commercials: a high-resolution EEG study. IEEE Transactions on Neural Systems and Rehabilitation Engineering: A Publication of the IEEE Engineering in Medicine and Biology Society 16(6), 522–531 (2008c)

Astolfi, L., et al.: Comparison of different cortical connectivity estimators for high-resolution EEG recordings. Human Brain Mapping 28(2), 143–157 (2007a)

Astolfi, L., et al.: Imaging functional brain connectivity patterns from high-resolution EEG and fMRI via graph theory. Psychophysiology 44(6), 880–893 (2007b)

Babiloni, C., Carducci, F., Del Gratta, C., Demartin, M., Romani, G.L., Babiloni, F., Rossini, P.M.: Hemispherical asymmetry in human SMA during voluntary simple unilateral movements. An fMRI study. Cortex 39(2), 293–305 (2003)

Babiloni, C., et al.: Mapping of early and late human somatosensory evoked brain potentials to phasic galvanic painful stimulation. Human Brain Mapping 12(3), 168–179 (2001)

Babiloni, F., et al.: Estimation of the cortical functional connectivity with the multimodal integration of high-resolution EEG and fMRI data by directed transfer function. NeuroImage 24(1), 118–131 (2005)

Babiloni, F., et al.: High-resolution electro-encephalogram: source estimates of Laplacian-transformed somatosensory-evoked potentials using a realistic subject head model constructed from magnetic resonance images. Medical & Biological Engineering & Computing 38(5), 512–519 (2000)

Babiloni, F., et al.: High resolution EEG: a new model-dependent spatial deblurring method using a realistically-shaped MR-constructed subject's head model. Electroencephalography and Clinical Neurophysiology 102(2), 69–80 (1997)

Baccalà, L.A., Sameshima, K.: Partial Directed Coherence: a new concept in neural structure determination. Biol. Cybern. 84, 463–474 (2001)

Bai, X., et al.: Evaluation of cortical current density imaging methods using intracranial electrocorticograms and functional MRI. NeuroImage 35(2), 598–608 (2007)

Baldaro, B., Mazzetti, M., Codispoti, M., Tuozzi, G., Bolzani, R., Trombini, G.: Autonomic reactivity during viewing of an unpleasant film. Percept. Mot. Skills 93, 797–805 (2001)

Bargh, J.A., Chartrand, T.L.: The Unbearable Automaticity of Being. American Psychologist 54(7), 462–479 (1999)

Baumgartner, T., Esslen, M., Jancke, L.: From emotion perception to emotion experience: Emotions evoked by pictures and classical music. International Journal of Psychophysiology 60(1), 34–43 (2006)

Beckwith, N.E., Lehmann, D.R.: The Importance of Halo Effects in Multi-Attribute Attitude Models. Journal of Marketing Research 12(3), 265–275 (1975)

Benjamini, Y., Hochberg, Y.: Controlling the false discovery rate: a practical and powerful approach to multiple testing. Journal of the Royal Statistical Society, Series B (Methodological) 57, 125–133 (1995)

Benjamini, Y., Daniel, Y.: The control of the false discovery rate in multiple testing under dependency. Annals of Statistics 29(4), 1165–1188 (2001), doi:10.1214/aos/1013699998

Berntson, G.G., et al.: Heart rate variability: origins, methods, and interpretive caveats. Psychophysiology 34(6), 623–648 (1997)

Biener, L., Harris, J.E., Hamilton, W.: Impact of the Massachusetts tobacco control programme: Population based trend analysis. British Medical Journal 321, 351–354 (2000)

Blood, A.J., et al.: Emotional responses to pleasant and unpleasant music correlate with activity in paralimbic brain regions. Nature Neuroscience 2(4), 382–387 (1999)

Bonferroni, C.: Teoria statistica delle classi e calcolo delle probabilità. Pubblicazioni del R Istituto Superiore di Scienze Economiche e Commerciali di Firenze 8, 3–62 (1936)

Boucsein, W.: Electrodermal Activity, 1st edn. Springer (1992)

Braeutigam, S.: Neuroeconomics–from neural systems to economic behaviour. Brain Research Bulletin 67(5), 355–360 (2005)

Braeutigam, S., Rose, S.P., Swithenby, S.J., Ambler, T.: The distributed neuronal systems supporting choice-making in real-life situations: differences between men and women when choosing groceries detected using magne-toencephalography. The European Journal of Neuroscience 20(1), 293–302 (2004)

Braeutigam, S., Stins, J.F., Rose, S.P., Swithenby, S.J., Ambler, T.: Magnetoencephalographic signals identify stages in real-life decision processes. Neural Plasticity 8(4), 241–254 (2001)

Breiter, H.C., et al.: Functional imaging of neural responses to expectancy and experience of monetary gains and losses. Neuron 30(2), 619–639 (2001)

Breiter, H.C., et al.: Acute effects of cocaine on human brain activity and emotion. Neuron 19(3), 591–611 (1997)

Brett, M., Penny, W., Kiebel, S.: An introduction to Random Field Theory. Academic Press, London (2004)
Brothers, L.: The social brain: A project for integrating primate behavior and neurophysiology in a new domain. Concepts in Neuroscience 1, 27–51
Buck, R.: The biological affects: A typology. Psycholog Rev. 106, 301–336 (1999)
Cacioppo, J.T., et al.: Foundations in Social Neuroscience, illustrated edn. The MIT Press (2002)
Cahill, L., Haier, R.J., Fallon, J., Alkire, M.T., Tang, C., Keator, D., Wu, J., McGaugh, J.L.: Amygdala activity at encoding correlated with long-term, free recall of emotional information. Proceedings of the National Academy of Sciences of the United States of America 93(15), 8016–8021 (1996)
Coan, J.A., Schaefer, H.S., Davidson, R.J.: Lending a hand: social regulation of the neural response to threat. Psychological Science 17(12), 1032–1039 (2006)
Critchley, H.D.: Electrodermal responses: what happens in the brain. The Neuroscientist: A Review Journal Bringing Neurobiology, Neurology and Psychiatry 8(2), 132–142 (2002)
Dale, A.M., et al.: Dynamic statistical parametric mapping: combining fMRI and MEG for high-resolution imaging of cortical activity. Neuron 26(1), 55–67 (2000)
Dale, A.M., Sereno, M.: Improved localization of cortical activity by combining EEG and MEG with MRI cortical surface reconstruction: a linear approach. J. Cognitive Neuroscience 5, 162–176 (1993)
Damasio, A.R., Damasio, H., Christen, Y.: Neurobiology of Decision-Making, 1st edn. Springer (1996)
Damasio, A.R.: Emotion in the perspective of an integrated nervous system. Brain Res. Rev. 26, 83–86 (1998)
Darwin, C.: The Expression of the Emotions in Man and Animals. Univ. Chicago Press, Chicago (1872)
Davidson, R.J.: What does the prefrontal cortex do" in affect: perspec-tives on frontal EEG asymmetry research. Biological Psychology 67(1-2), 219–233 (2004)
Davidson, R.J.: Anxiety and affective style: role of prefrontal cortex and amygdala. Biological Psychiatry 51(1), 68–80 (2002)
Davidson, R.J.: Affective style, psychopathology, and resilience: brain mechanisms and plasticity. The American Psychologist 55(11), 1196–1214 (2000)
Davidson, Irwin: The functional neuroanatomy of emotion and affective style. Trends in Cognitive Sciences 3(1), 11–21 (1999)
Davidson, R.J., et al.: Regional brain function, emotion and disorders of emotion. Current Opinion in Neurobiology 9(2), 228–234 (1999)
Day, G.S.: The Threats to Marketing Research. Journal of Marketing Research 12(4), 462–467 (1975)
De Vico Fallani, F., et al.: Cortical functional connectivity networks in normal and spinal cord injured patients: Evaluation by graph analysis. Human Brain Mapping 28(12), 1334–1346 (2007)
Delgado, M.R., et al.: Tracking the hemodynamic responses to reward and punishment in the striatum. Journal of Neurophysiology 84(6), 3072–3077 (2000)
Deppe, M., et al.: Nonlinear responses within the medial prefrontal cortex reveal when specific implicit information influences economic decision making. Journal of Neuroimaging: Official Journal of the American Society of Neuroimaging 15(2), 171–182 (2005)
Ding, L., Lai, Y., He, B.: Low resolution brain electromagnetic tomo-graphy in a realistic geometry head model: a simulation study. Physics in Medicine and Biology 50(1), 45–56 (2005)

Eisenberger, N.I., Lieberman, M.D., Williams, K.D.: Does rejection hurt? An FMRI study of social exclusion. Science 302(5643), 290–2 (2003)

Elliott, R., Friston, K.J., Dolan, R.J.: Dissociable neural responses in human reward systems. The Journal of Neuroscience: The Official Journal of the Society for Neuroscience 20(16), 6159–6165 (2000)

Emery, S., Wakefield, M., Terry-McElrath, Y., Szczypka, G., O'Malley, P.M., Johnston, L.D., Chaloupka, F.J., Flay, B.: Televised state sponsored anti-tobacco advertising and youth smoking beliefs and behavior in the United States, 1999–2000. Archives Pediatric Adolescent Medicine 159, 639–645 (2005)

Erk, S., et al.: Cultural objects modulate reward circuitry. Neuroreport 13(18), 2499–2503 (2002)

Fallani, F.D.V., Rodrigues, F.A., Fontoura Costa, L., Astolfi, L., Cincotti, F., Mattia, D., Salinari, S., Babiloni, F.: Multiple Pathways Analysis of Brain Functional Networks from EEG Signals: An Application to Real Data. Brain Topography 23, 344–354 (2011)

Fowles, D.C., et al.: Committee report. Publication recommendations for electrodermal measurements. Psychophysiology 18(3), 232–239 (1981)

Frank, L.R.: Ventromedial prefrontal cortex activation is critical for preference judgments. Neuroreport 14, 1311–1315 (2003)

Franzen, G., Bouwman, M.: The Mental World of Brands: Mind, Memory and Brand Success. NTC Publications (2001)

Friston, K.J.: Functional and effective connectivity in neuroimaging: a synthesis. Human Brain Mapping 2, 56–78 (1994)

Friston, K.J., Harrison, L., Penny, W.: Dynamic causal modelling. Neuroimage 19, 1273–1302 (2003)

Genovese, C.R., Lazar, N.A., Nichols, T.: Thresholding of statistical maps in functional neuroimaging using the false discovery rate. Neuroimage 15(4), 870–878 (2002)

Gevins, A., et al.: High resolution EEG: 124-channel recording, spatial deblurring and MRI integration methods. Electroencephalography and Clinical Neurophysiology 90(5), 337–358 (1994)

Gottfried, J.A., O'Doherty, J., Dolan, R.J.: Encoding predictive reward value in human amygdala and orbitofrontal cortex. Science 301(5636), 1104–1107 (2003)

Gottfried, J.A., O'Doherty, J., Dolan, R.J.: Appetitive and Aversive Ol-factory Learning in Humans Studied Using Event-Related Functional Magnetic Resonance Imaging. J. Neurosci. 22(24), 10829–10837 (2002)

Grabenhorst, F., Rolls, E.T., Parris, B.A.: From affective value to decision-making in the prefrontal cortex. The European Journal of Neuroscience 28(9), 1930–1939 (2008)

Grave de Peralta Menendez, R., Gonzalez Andino, S.L.: Distributed source models: standard solutions and new developments. In: Uhl, C. (ed.) Analysis of Neurophysiological Brain Functioning, pp. 176–201. Springer (1998)

Grave de Peralta, R., Hauk, O., Gonzalez Andino, S., Vogt, H., Michel, C.M.: Linear inverse solution with optimal resolution kernels applied to the electromagnetic tomography. Human Brain Mapping 5, 454–467 (1997)

Grave de Peralta Menendez, R., Murray, M.M., Michel, C.M., Martuzzi, R., Gonzalez Andino, S.L.: Electrical neuroimaging based on biophysical constraints. Neuroimage 21(2), 527–539 (2004)

Green, P.E., Srinivasan, V.: Conjoint Analysis in Marketing: New Developments with Implications for Research and Practice. The Journal of Marketing 54(4), 3–19 (1990)

Griffin, A., Hauser, J.R.: The Voice of the Customer. Marketing Science 12(1), 1–27 (1993)

Hallez, H., Vanrumste, B., Grech, R., Muscat, J., De Clercq, W., Vergult, A., D'Asseler, Y., Camilleri, K.P., Fabri, S.G., Van Huffel, S., Lemahieu, I.: Review on solving the forward problem in EEG source analysis. J. Neuroeng. Rehabil. 30, 4–46 (2007)

Hämäläinen, M., Ilmoniemi, R.: Interpreting measured magnetic field of the brain: Estimates of the current distributions. Technical report TKK-F-A559, Helsinki University of Technology (1984)

Handy, T.C., Smilek, D., Geiger, L., Liu, C., Schooler, J.W.: ERP evidence for rapid hedonic evaluation of logos. Journal of Cognitive Neuroscience 22(1), 124–138 (2010)

Hastie, R.: Causes and effects of causal attributions. J. Personal Soc. Psychol. 46, 44–56 (1984)

He, B., Hori, J., Babiloni, F.: EEG Inverse Problems. In: Akay, M. (ed.) Wiley Encyclopedia in Biomedical Engineering, pp. 1355–1363. John Wiley & Sons, Inc. (2006)

He, B., Wang, Y., Wu, D.: Estimating cortical potentials from scalp EEG's in a realistically shaped inhomogeneous head model by means of the boundary element method. IEEE Transactions on BioMedical Engineering 46(10), 1264–1268 (1999)

Hommer, D.W., et al.: Amygdalar recruitment during anticipation of monetary rewards: an event-related fMRI study. Annals of the New York Academy of Sciences 985, 476–478 (2003)

Horwitz, B.: The elusive concept of brain connectivity. Neuroimage 19, 466–470 (2003)

Ioannides, A.A.: Magnetoencephalography as a research tool in neuroscience: State of the art. Neuroscientist 12, 524–544 (2006)

Ioannides, A.A., et al.: Real time processing of affective and cognitive stimuli in the human brain extracted from MEG signals. Brain Topography 13(1), 11–19 (2000)

Kato, J., et al.: Neural correlates of attitude change following positive and negative advertisements. Frontiers in Behavioral Neuroscience 3, 6 (2009)

Kaminski, M.J., Blinowska, K.J.: A new method of the description of the information flow in the brain structures. Biol. Cybern. 65(3), 203–210 (1991)

Kaminski, M., et al.: Evaluating causal relations in neural systems: granger causality,directed transfer function and statistical assessment of significance. Biol. Cybern. 85(2), 145–157 (2001)

Kimura, D.: Sex, sexual orientation and sex hormones influence human cognitive function. Current Opinion in Neurobiology 6(2), 259–263 (1996)

Klimesch, W.: EEG alpha and theta oscillations reflect cognitive and memory performance: a review and analysis. Brain Res. Rev. 29(2), 169–195 (1999)

Klucharev, V., Smidts, A., Fernández, G.: Brain mechanisms of persua-sion: how 'expert power' modulates memory and attitudes. Social Cognitive and Affective Neuroscience 3(4), 353–366 (2008)

Knutson, B., et al.: Neural predictors of purchases. Neuron 53(1), 147–156 (2007)

Knutson, B., et al.: A region of mesial prefrontal cortex tracks monetar-ily rewarding outcomes: characterization with rapid event-related fMRI. NeuroImage 18(2), 263–272 (2003)

Knutson, B., Adams, C.M., et al.: Anticipation of increasing monetary reward selectively recruits nucleus accumbens. The Journal of Neuroscience: The Official Journal of the Society for Neuroscience 21(16), RC159 (2001a)

Knutson, B., Fong, G.W., et al.: Dissociation of reward anticipation and outcome with event-related fMRI. Neuroreport 12(17), 3683–3687 (2001b)

Knutson, B., et al.: FMRI visualization of brain activity during a monetary incentive delay task. NeuroImage 12(1), 20–27 (2000)

Koenigs, M., Young, L., Adolphs, R., et al.: Damage to the prefrontal cortex increases utilitarian moral judgments. Nature 446, 908–911 (2007)

Kreibig, S.D., et al.: Cardiovascular, electrodermal, and respiratory response patterns to fear- and sadness-inducing films. Psychophysiology 44(5), 787–806 (2007)

Kringelbach, M.: Food for thought: hedonic experience beyond Homeostasis in the human brain. Neuroscience 126, 807–819 (2004)

Kringelbach, M.L., Rolls, E.T.: The functional neuroanatomy of the human orbitofrontal cortex: evidence from neuroimaging and neuropsychology. Progress in Neurobiology 72(5), 341–372 (2004)

Kringelbach, M.L., O'Doherty, J., Rolls, E.T., Andrews, C.: Activation of the human orbitofrontal cortex to a liquid food stimulus is correlated with its subjective pleasantness. Cereb Cortex 3, 1064–1071 (2003)

Lamm, C., Batson, C.D., Decety, J.: The neural substrate of human em-pathy: effects of perspective-taking and cognitive appraisal. Journal of Cognitive Neuroscience 19, 42–58 (2007)

Langleben, D.D., Loughead, J.W., Ruparel, K., Hakun, J.G., Busch-Winokur, S., Holloway, M.B., Strasser, A.A., Cappella, J.N., Lerman, C.: Reduced prefrontal and temporal processing and recall of high "sensation value" ads. Neuroimage 15, 46(1), 219–225 (2009)

Latora, V., et al.: Efficient behavior of small-world networks. Phys. Rev. Lett. 87, 198701 (2001)

Lawson, C.L., Hanson, R.J.: Solving Least Squares Problems. Prentice Hall, Englewood Cliffs (1974)

Le, J., Gevins, A.: Method to reduce blur distortion from EEG's using a realistic head model. IEEE Transactions on Bio-Medical Engineering 40(6), 517–528 (1993)

LeDoux, J.: The emotional brain, fear, and the amygdala. Cellular and Molecular Neurobiology 23(4-5), 727–738 (2003)

Lehmann, D., Skrandies, W.: Reference-free identification of compo-nents of checkerboard-evoked multichannel potential fields. Electroencephalography and Clinical Neurophysiology 48(6), 609–621 (1980)

Logan, B.R., Geliazkova, M.P., Rowe, D.B.: An evaluation of spatial threshold-ing techniques in fMRI analysis. Hum. Brain Mapp. 29(12), 1379–1389 (2008)

Lieberman, M.D., Gaunt, R., Gilbert, D.T., et al.: Reflection and reflexion: A social cognitive neuroscience approach to attributional inference. In: Zanna, M. (ed.) Advances in Experimental Social Psychology, pp. 199–249. Academic Press, New York (2002)

Liu, A.K.: Spatiotemporal brain imaging, PhD dissertation, Massachusetts Institute of Technology, Cambridge, MA (2000)

Maddock, R.J.: The retrosplenial cortex and emotion: new insights from functional neuroimaging of the human brain. Trends in Neurosciences 22(7), 310–316 (1999)

Malik, et al.: Heart rate variability. Standards of measurement, physiological interpretation, and clinical use. Task Force of the European Society of Cardiology and the North American Society of Pacing and Electrophysiology. European Heart Journal 17(3), 354–381 (1996)

Malliani, A.: Heart rate variability: from bench to bedside. European Journal of Internal Medicine 16(1), 12–20 (2005)

Manuck, S.B., Flory, J., Muldoon, M., et al.: Is there a neurobiology of in-tertemporal choice. In: Loewenstein, G.F., Read, D., Baumeister, R. (eds.) Time and Decision: Economic and Psychological Perspectives on Intertemporal Choice. Russell Sage, New York (2003)

McCabe, K., et al.: A functional imaging study of cooperation in two-person reciprocal exchange. Proceedings of the National Academy of Sciences of the United States of America 98(20), 11832–11835 (2001)

McClure, S.M., et al.: Neural correlates of behavioral preference for culturally familiar drinks. Neuron 44(2), 379–387 (2004)

McDonald, C.: Is Your Advertising Working? WARC (2003)

Mclntosh, A.R., Gonzalez-Lima, F.: Structural equation modeling and its application to network analysis in functional brain imaging. Human Brain Mapping 2, 2–22 (1994)

Mendez, M., et al.: Time-varying analysis of the heart rate variability during REM and non REM sleep stages. In: Conference Proceedings:. Annual International Conference of the IEEE Engineering in Medicine and Biology Society. IEEE Engineering in Medicine and Biology Society. Conference, vol. 1, pp. 3576–3579 (2006)

Menke, W.: Geophysical Data Analysis: Discrete Inverse Theory. Academic Press, San Diego (1989)

Michel, C.M., Murray, M.: Towards the utilization of EEG as a brain im-aging tool. Neuroimage 61(2), 371–385 (2012), doi:10.1016/j.neuroimage.2011.12.039

Michel, C.M., Murray, M.M., Lantz, G., Gonzalez, S., Spinelli, L., Grave de Peralta, R.: EEG source imaging. Clin Neurophysiol. 115(10), 2195-222 (2004)

Montano, N., et al.: Heart rate variability explored in the frequency domain: a tool to investigate the link between heart and behavior. Neuroscience and Biobehavioral Reviews 33(2), 71–80 (2009)

Morris, J.D., et al.: Mapping a multidimensional emotion in response to television commercials. Human Brain Mapping 30(3), 789–796 (2009)

Nagai, Y., et al.: Brain activity relating to the contingent negative variation: an fMRI investigation. NeuroImage 21(4), 1232–1241 (2004)

Nichols, T., Hayasaka, S.: ontrolling the familywise error rate in func-tional neuroimaging: a comparative review. Stat. Methods Med. Res. 12(5), 419–446 (2003)

Nunez, P.L.: Neocortical Dynamics and Human EEG Rhythms, 1st edn. Oxford University Press, USA (1995)

O'Doherty, J., et al.: Dissociating valence of outcome from behavioral control in human orbital and ventral prefrontal cortices. The Journal of Neuroscience: The Official Journal of the Society for Neuroscience 23(21), 7931–7939 (2003)

O'Doherty, J.P., et al.: Neural responses during anticipation of a primary taste reward. Neuron 33(5), 815–826 (2002)

O'Doherty, J., et al.: Abstract reward and punishment representations in the human orbitofrontal cortex. Nature Neuroscience 4(1), 95–102 (2001)

Oliveri, M., Babiloni, C., Filippi, M.M., Caltagirone, C., Babiloni, F., Cicinelli, P., Traversa, R., Palmieri, M.G., Rossini, P.M.: Influence of the supplementary motor area on primary motor cortex excitability during movements triggered by neutral or emotionally unpleasant visual cues. Exp Brain Res. 149(2), 214–21 (2003)

Palomba, D., Sarlo, M., Angrilli, A., Mini, A., Stegagno, L.: Cardiac re-sponses associated with affective processing of unpleasant film stimuli. Int. J. Psychophysiol. 36, 45–57 (2000)

Panksepp, J.: Affective Neuroscience: The Foundations of Human and Animal Emotions, 1st edn. Oxford University Press, USA (2004)

Pascual-Marqui, R.D.: Reply to comments by Hamalainen, Ilmoniemi and Nunez. In: Skrandies, W. (ed.) ISBET Newsletter, vol. (6), pp. 16–28 (December 1995)

Pascual-Marqui, R.D.: Standardized low-resolution brain electromagnetic tomography (sLORETA): technical details. Methods Find Exp. Clin. Pharmacol. 24(suppl. D), 5–12 (2002)

Ramnani, N., Miall, R.C.: Instructed delay activity in the human pre-frontal cortex is modulated by monetary reward expectation. Cerebral Cortex 13(3), 318–327 (2003)

Rao, C.R., Mitra, S.K.: Generalized inverse of matrices and its applications. Wiley, New York (1977)

Ribary, U., Ioannides, A.A., Singh, K.D., Hasson, R., Bolton, J.P.R., Lado, F., Mogilner, A., Llinas, R.: Magnetic-Field Tomography of Coherent Tha-lamocortical 40-Hz Oscillations in Humans. Proc. Natl. Acad. Sci.USA 88, 11037–11041 (1991)

Rilling, J.K., et al.: Opposing BOLD responses to reciprocated and unre-ciprocated altruism in putative reward pathways. Neuroreport 15(16), 2539–2543 (2004a)

Rilling, J.K., et al.: The neural correlates of theory of mind within in-terpersonal interactions. NeuroImage 22(4), 1694–1703 (2004b)

Rilling, J., et al.: A neural basis for social cooperation. Neuron 35(2), 395–405 (2002)

Rolls, E.T., et al.: Representations of pleasant and painful touch in the human orbitofrontal and cingulate cortices. Cerebral Cortex 13(3), 308–317 (2003)

Rolls, E.T.: The orbitofrontal cortex and reward. Cerebral Cortex 10(3), 284–294 (2000)

Rossiter, J.R., Silberstein, R.B., Harris, P.G., Nield, G.A.: Brain-imaging detection of visual scene encoding in long-term memory for TV commercials (2001)

Sanfey, A.G., et al.: The neural basis of economic decision-making in the Ultimatum Game. Science 300(5626), 1755–1758 (2003)

Stam, C.J.: Functional connectivity patterns of human magnetoencephalographic recordings: a 'small-world' network? Neurosci. Lett. 355, 25–28 (2004)

Stam, C., et al.: Small-World Networks and Functional Connectivity in Alzheimer's Disease. Cereb. Cortex (2006)

Stefan, S., Harald, W.: Electrodermal Activity. State Of The Art Measurement And Techniques For Parapsychological Purposes - EDA (2000)

Schneider, W., Shiffrin, R.M.: Controlled and automatic human informa-tion processing: I. Detection, search and attention. Psycholog Rev. 1, 1–66 (1977)

Schultz, W.: Neural coding of basic reward terms of animal learning theory, game theory, microeconomics and behavioural ecology. Current Opinion in Neurobiology 14(2), 139–147 (2004)

Shadel, W., Niaura, R., Abrams, B.: Adolescents' reactions to the im-agery displayed in smoking and antismoking advertisements. Psychology of Addictive Behaviors 16, 173–176 (2002)

Silberstein, R.B., et al.: Frontal steady-state potential changes predict long-term recognition memory performance. International Journal of Psychophysiology: Official Journal of the International Organization of Psychophysiology 39(1), 79–85 (2000)

Singer, T., et al.: Brain responses to the acquired moral status of faces. Neuron 41(4), 653–662 (2004)

Small, D.M., et al.: Dissociation of neural representation of intensity and affective valuation in human gustation. Neuron 39(4), 701–711 (2003)

Small, D.M., et al.: Changes in brain activity related to eating chocolate: from pleasure to aversion. Brain: A Journal of Neurology 124(pt. 9), 1720–1733 (2001)

Somerville, L.H., Heatherton, T.F., Kelley, W.M.: Anterior cingulate cortex responds differentially to expectancy violation and social rejection. Nature Neuroscience 9, 1007–1008 (2006)

Spiegel, M.: Theory and problems of vector analysis and an introduction to tensor analysis. Mc Graw Hill, New York (1978)

Spitzer, M.: Models of schizophrenia: from neuroplasticity and dopamine to psychopathology and clinical management. In: Lecrubier, K.S.Y. (ed.) Dopamine in the Pathophysiology and Treatment of Schizophrenia, Martin Dunitz, pp. 155–175. Taylor & Francis Group, London (2003)

Sporns, O., Zwi, J.D.: The small world of the cerebral cortex. Neuroinformatics 2, 145–162 (2004)

Steptoe, A., Wardle, J.: Emotional fainting and the psychophysiologic response to blood and injury: autonomic mechanisms and coping strategies. Psychosom. Med. 50, 402–417 (1988)

Summerfield, C., Mangels, J.A.: Coherent theta-band EEG activity predicts item-context binding during encoding. NeuroImage 24(3), 692–703 (2005)

Strogatz, S.H.: Exploring complex networks. Nature 410, 268–276 (2001)

Thut, G., Schultz, W., Roelcke, U., et al.: Activation of the human brain by monetary reward. Neuroreport 8, 1225–1228 (1997)

Tichonov, A.N., Arsenin, V.Y.: Solutions of ill-posed problems. Winston, Washington, D.C. (1977)

Tononi, G., Sporns, O.: Measuring information integration. BMC Neuroscience 4, 31 (2003)

Tulving, E., Kapur, S., Craik, F.I., Moscovitch, M., Houle, S.: Hemispheric encoding/retrieval asymmetry in episodic memory: positron emission tomo-graphy findings. Proceedings of the National Academy of Sciences of the United States of America 91(6), 2016–2020 (1994)

Ullsperger, M., von Cramon, D.Y.: Decision making, performance and out-come monitoring in frontal cortical areas. Nature Neurosci. 7, 1173–1174 (2004)

Urbano, A., et al.: Dynamic functional coupling of high resolution EEG potentials related to unilateral internally triggered one-digit movements. Electroencephalography and Clinical Neurophysiology 106(6), 477–487 (1998)

Vecchiato, G., Toppi, J., Astolfi, L., Cincotti, F., De Vico Fallani, F., Maglione, A.G., Borghini, G., Cherubino, P., Mattia, D., Babiloni, F.: The added value of the electrical neuroimaging for the evaluation of marketing stimuli. Bulletin of the Polish Academy of Sciences Technical Sciences 60(3) (2012a)

Vecchiato, G., Maglione, A.G., Cherubino, P., Kong, W., Hu, S., Wei, D., Colosimo, A., Babiloni, F.: Increase of theta EEG activity correlates with memorization of TV commercials for Chinese and Italian subjects. In: Proceedings of the 7th International Workshop on Biosignal Interpretation. Como, July 2-4 (2012b)

Vecchiato, G., Maglione, A., Kong, W., Hu, S., Wei, D., Colosimo, A., Babiloni, F.: Differences in emotional processing during the observation of Western and Eastern TV commercials measured by EEG in a group of Eastern subjects. In: Conference Proceedings of IFMBE 2012, Beijing, May 28-31 (2012c)

Vecchiato, G., Kong, W., Maglione, A.G., Wei, D.: Understanding the Impact of TV Commercials: Electrical Neuroimaging. IEEE Pulse 3(3), 42–47 (2012d)

Vecchiato, G., Astolfi, L., De Vico Fallani, F., Toppi, J., Aloise, F., Bez, F., Wei, D., Kong, W., Dai, J., Cincotti, F., Mattia, D., Babiloni, F.: On the Use of EEG or MEG Brain Imaging Tools in Neuromarketing Research. Computational Intelligence and Neuroscience 2011, Article ID 643489, 12 pages (2011a)

Vecchiato, G., Toppi, J., Astolfi, L., De Vico Fallani, F., Cincotti, F., Mattia, D., et al.: Spectral EEG Frontal Asymmetries Correlate with the Experienced Pleasantness of TV Commercial Advertisements. Medical & Biological Engineering & Computing (2011b)

Vecchiato, G., De Vico Fallani, F., Astolfi, L., Toppi, J., Cincotti, F., Mattia, D., Salinari, S., Babiloni, F.: The issue of multiple univariate comparisons in the context of neuroelectric brain mapping: an application in a neuromarketing experiment. Journal of Neuroscience Methods 191(2), 283–289 (2010a)

Vecchiato, G., Astolfi, L., De Vico Fallani, F., Cincotti, F., Mattia, D., Salinari, S., Soranzo, R., Babiloni, F.: Changes in brain activity during the observa-tion of TV commercials by using EEG, GSR and HR measurements. Brain Topogr. 23(2), 165–179 (2010b)

Vecchiato, G., Astolfi, L., Cincotti, F., De Vico Fallani, F., Sorrentino, D.M., Mattia, D., Salinari, S., et al.: Patterns of cortical activity during the observation of Public Service Announcements and commercial advertisings. Nonlinear Biomedical Physics 4(suppl. 1), S3 (2010c)

Venables, P.H.: Autonomic activity. Annals of the New York Academy of Sciences 620, 191–207 (1991)

Wakefield, M., Flay, B., Nichter, M., Giovino, G.: Effects of anti-smoking advertis-ing on youth smoking: a review. Journal of Health Communication 8, 229–247 (2003)

Walter, H., Abler, B., Ciaramidaro, A., Erk, S.: Motivating forces of human actions. Neuroimaging reward and social interaction. Brain Res Bull. 67(5), 368–381 (2005)

Walter, H.I.: Funktionelle Bildgebung in Psychiatrie und Psychotherapie. Schattauer GmbH (2004)

Welch, P.D.: The Use of Fast Fourier Transform for the Estimation of Power Spectra: A Method Based on Time Averaging Over Short. Modified Periodograms 15, 70–73 (1967)

Werkle-Bergner, M., et al.: Cortical EEG correlates of successful mem-ory encoding: implications for lifespan comparisons. Neuroscience and Biobehavioral Reviews 30(6), 839–854 (2006)

Young, C.: Brain Waves, Picture Sorts®, and Branding Moments. J. Advert. Res., 42–53 (2002)

Zajonc, R.: Emotions. In: Gilbert, D., Fiske, S., Lindzey, G. (eds.) The Handbook of Social Psychology, pp. 591–632. Oxford University Press, New York (1998)

Zaltman, G.: How Customers Think: Essential Insights into the Mind of the Market, 1st edn. Harvard Business Press (2003)

Zar, J.H.: Biostatistical Analysis, 5th edn. Prentice Hall, Englewood Cliffs (2009)